Adaptive Dynamic Programming

Jiayue Sun · Shun Xu · Yang Liu · Huaguang Zhang

Adaptive Dynamic Programming

For Chemotherapy Drug Delivery

Springer

Jiayue Sun
The State Key Laboratory of Synthetical Automation for Process Industries and the College of Information Science and Engineering
Northeastern University
Shenyang, Liaoning, China

Yang Liu
The Department of Thoracic Surgery
The First Affiliated Hospital of China Medical University
Shenyang, Liaoning, China

Shun Xu
The Department of Thoracic Surgery
The First Affiliated Hospital of China Medical University
Shenyang, Liaoning, China

Huaguang Zhang
The State Key Laboratory of Synthetical Automation for Process Industries and the College of Information Science and Engineering
Northeastern University
Shenyang, Liaoning, China

ISBN 978-981-99-5931-0 ISBN 978-981-99-5929-7 (eBook)
https://doi.org/10.1007/978-981-99-5929-7

This work was supported by China National Postdoctoral Program for Innovative Talents (BX20220060) and National Natural Science Foundation of China (62203469, 62203097)

This Springer imprint is published by the registered company Springer Nature Singapore Pte Ltd.
The registered company address is: 152 Beach Road, #21-01/04 Gateway East, Singapore 189721, Singapore

Paper in this product is recyclable.

To My Family

Jiayue Sun

To My Family

Shun Xu

To My Family

Yang Liu

To My Family

Huaguang Zhang

Preface

Optimization is the process of finding the best solution to a problem subject to a set of constraints. It has long been a cornerstone of both engineering and mathematics. The evolution of optimization can be traced back to the ancient Greeks, who employed geometric methods to solve optimization problems. In the eighteenth and nineteenth centuries, optimization began to take on a more formalized approach with the development of calculus and the rise of industrialization. Mathematicians such as Leonhard Euler and Joseph-Louis Lagrange developed methods for finding the maximum and minimum values of functions, which were crucial for optimizing industrial processes and designing efficient machines. The twentieth century saw a significant expansion in the field of optimization, with the development of linear programming and other optimization techniques. Linear programming is a mathematical technique for optimizing a linear objective function subject to linear constraints, which involves optimizing a linear objective function subject to linear constraints, was first introduced by George Dantzig in the 1940s and quickly became a powerful tool for solving complex optimization problems. In the latter half of the twentieth century, optimization began to be applied to a wide range of fields beyond mathematics and engineering. Operations research, which uses mathematical models to optimize complex systems, became a popular field in business and management. Optimization techniques were also applied to fields such as finance, transportation, and telecommunications. These models are typically static and deterministic, meaning they do not take into account the dynamic nature of many real-world systems. Over time, researchers have developed increasingly sophisticated optimization algorithms that can adapt to changing conditions and learn from experience.

Adaptive dynamic programming (ADP) is a powerful optimization technique for improving dynamic systems. Despite being a relatively new area of optimization, it has received broad usage in various industries. ADP is rooted in the principle of dynamic programming, which involves breaking down a complex optimization problem into smaller subproblems. These subproblems are solved recursively using backward induction. ADP learns from feedback and adjusts its behavior accordingly, making it useful for systems operating in uncertain environments. It has been applied to a wide range of problems, including electrical power systems, robotics, self-driving cars, and trading strategies. This book focuses on the practical application of ADP in chemotherapy drug delivery, taking into account clinical variables and real time data. ADP's ability to adapt to changing conditions and make optimal decisions in complex and uncertain situations makes it a valuable tool in addressing pressing challenges in healthcare and other fields. As optimization technology evolves, we can expect to see even more sophisticated and powerful solutions emerge.

Shenyang, China
April 2023

Jiayue Sun
Shun Xu
Yang Liu
Huaguang Zhang

Acknowledgements Our acknowledgments also go to our fellow colleagues who have offered invaluable support and encouragement throughout this research effort. The authors are especially grateful to their families for their encouragement and never-ending support when it was most required. Finally, we would like to thank the editors at Springer for their professional and efficient handling of this project.

The writing of this book was partially supported by China National Postdoctoral Program for Innovative Talents (BX20220060), National Natural Science Foundation of China (62203097, 62203469).

Contents

List of Figures

List of Tables

Chapter 1
Introduction

The stability analysis of dynamical systems, which are ubiquitous in nature, has long been a hot topic of research and several approaches have been proposed. However, control scientists often demand optimality in addition to the stability of the control system. In the 1950s and 1960s, motivated by the development of space technology and the practical use of digital computers, the theory of optimization of dynamical systems developed rapidly, forming an important branch of the discipline: optimal control. It is increasingly used in many fields, such as space technology, systems engineering, economic management and decision-making, population control, and optimization of multi-stage process equipment. In 1957, Bellman proposed an effective tool for solving optimal control problems: the dynamic programming (DP) method [1]. At the heart of this approach is Bellman's optimality principle, which states that the optimal policy for a multilevel decision process has the property that, regardless of the initial state and initial decision, the remaining decisions must also be optimal for the state formed by the initial decision. This principle can be reduced to a basic recursive formula for solving multilevel decision problems by starting at the end and working backward to the beginning. It applies to a wide range of discrete, continuous, linear, nonlinear, deterministic, and stochastic systems.

ADP is a new approach to approximate optimality in the field of optimal control, and it is a current research topic in the international optimization field. The ADP method uses the function approximation structure to approximate the solution of the Hamilton-Jacobi-Bellman (HJB) equation and uses offline iteration or online update to obtain the approximate optimal control strategy of the system, which can effectively solve the optimal control problem of nonlinear systems [2–11]. Bertsekas et al. summarized neuronal dynamic programming in the literature [12, 13], describing in detail dynamic programming, the structure of neural networks, and training algorithms. Meanwhile, several effective methods have been proposed for applying neuronal dynamic programming. Si et al. summarized the development of ADP methods in cross-cutting disciplines and discussed the connection of DP and ADP methods with artificial intelligence, approximation theory, control theory, operations

J. Sun et al., *Adaptive Dynamic Programming*,
https://doi.org/10.1007/978-981-99-5929-7_1

research, and statistics [14]. In [15], Powell showed how to use ADP methods to solve deterministic or stochastic optimization problems, and pointed out the direction of ADP methods. In [16], Balakrishnan et al. concluded previous approaches to the design of feedback controllers for dynamic systems using the ADP method from both model and model-free cases. In [17], the ADP method was described from the perspective of requiring initial stability and not requiring initial stability.

The ADP method has a unique algorithm and structure compared to other existing optimal control methods. It overcomes the drawback that classical variational theory cannot handle optimal control problems with closed-set constraints on the control variables. Like the maximum value principle, the ADP method is not only suitable for optimal control problems with open-set constraints, but also for optimal control problems with closed-set constraints. While the extreme value principle can only provide the necessary conditions for optimal control problems, the DP method gives sufficient conditions. However, the direct application of the DP method is difficult due to the difficulty of solving the problem of "dimensional disaster" by the HJB equation in the DP method. Hence the ADP method, as an approximate solution to the DP method, overcomes the limitations of the DP method. It is more suitable for applications in systems with strong coupling, strong nonlinearity and high complexity. For example, the literature [18] presented a constrained adaptive dynamic programming (CADP) algorithm that could be used to solve general nonlinear non-affine optimal control problems with known dynamics. Unlike previous ADP algorithms, it was able to handle problems with state constraints directly by proposing a constrained generalized policy iteration framework that transforms the traditional policy improvement process into a constrained policy optimization problem with state constraints. To solve the problem of robust tracking control, the literature [19] designed an online adaptive learning structure to build a robust tracking controller for nonlinear uncertain systems. The literature [20] proposed an iterative method of bias policy for solving data-driven optimal control problems for unknown continuous linear systems by adding a bias parameter that could further relax the conditions of the initial admissible controller. The literature [21] considered the first attempt at ADP control for a nonlinear Itô-type stochastic system, which transformed a complex optimal tracking control problem into a stable control optimization problem by reconstructing a new stochastic augmented system. The use of a critical neural network in iterative learning subsequently simplifies the structure of the behavioral criterion and reduces the computational load. The ADP approach is still very widely used for a number of common practical systems. The literature [22] developed an event-triggered adaptive dynamic planning method to design formation controllers, and solved the problem of distributed formation control for multi-rotor UAS. For wind/light energy hybrid systems, literature [23] presented an adaptive dynamic programming method based on Bellman's principle, which enables accurate current sharing and voltage regulation. Based on this approach, it is possible to obtain the optimal control variables for each energy body objective.

Optimal control of nonlinear systems has been one of the hot spots and difficulties in the field of control research. As a novel technology to solve the optimal control problem, ADP method integrates the theories of neural network, adaptive evaluation design, augmented learning and classical dynamic programming, to overcome the problem of "dimensional disaster", which also enables the acquisition of an approximate optimal closed-loop feedback control law. As a consequence, delving deeper into the theory of ADP and its algorithms for solving optimal control of nonlinear systems holds immense theoretical significance and practical application value. Although the researches on the ADP method are still in its early stages, this book aims to equip readers with a foundational understanding of the method and empower them to apply it to diverse optimization problems in fields such as medicine, science, and engineering.

1.1 Optimal Control Formulation

There are several schemes of dynamic programming [1, 13, 24]. One can consider discrete-time systems or continuous-time systems, linear systems or nonlinear systems, time-invariant systems or time-varying systems, deterministic systems or stochastic systems, etc. Discrete-time (deterministic) nonlinear (time-invariant) dynamical systems will be discussed first. Time-invariant nonlinear systems cover most of the application areas and discrete time is the basic consideration for digital implementation.

1.1.1 ADP for Discrete-Time Systems

Consider the following discrete-time nonlinear systems:

$$x_{k+1} = F(x_k, u_k), k = 1, 2, ..., \tag{1.1}$$

where $x_k \in \mathbb{R}^n$ is the state vector and $u_k \in \mathbb{R}^m$ is the control input vector. The corresponding cost function (performance index function) of the system takes the form of

$$J(x_k, \overline{u}_k) = \sum_{i=k}^{\infty} \gamma^{i-k} U(x_i, u_i), \tag{1.2}$$

where $\overline{u}_k = (u_k, u_{k+1}, ...)$ is the control sequence starting at time k. $U(x_i, u_i)$ is the utility function. γ is the discount factor, meeting $0 < \gamma < 1$. Note that the function J is dependent on the initial time k and the initial state x_k. Generally, it is desired to determine $\overline{u}_0 = (u_0, u_1, ...)$ so that $J(x_0, \overline{u}_0)$ is optimized (i.e., maximized or minimized). We will use $\overline{u}_0^* = (u_0^*, u_1^*, ...)$ and $J^*(x_0)$ to denote the optimal control sequence and the optimal cost function, respectively.The objective of dynamic programming problem in this book is to determine a control sequence u_k, $k = 0, 1, ...,$

so that the function J (i.e., the cost) in (1.2) is minimized. The optimal cost function is defined as

$$J^*(x_0) = \inf_{\overline{u}_0} J(x_0, \overline{u}_0) = J(x_0, \overline{u}_0^*), \tag{1.3}$$

which is dependent upon the initial state x_0.

The control action may be determined as a function of the state. In this case, we write $u_k = u(x_k), \forall k$. Such a relationship, or mapping $u : R^n \to R^m$, is called feedback control, or control policy, or policy. It is also called control law. For a given control policy μ, the cost function in (1.2) is rewritten as

$$J^{\mu}(x_k) = \sum_{i=k}^{\infty} \gamma^{i-k} U(x_i, \mu(x_i)), \tag{1.4}$$

which is the cost function for system (1.1) starting at xk when the policy $u_k = \mu(x_k)$ is applied. The optimal cost for system (1.1) starting at x_0 is determined as

$$J^*(x_0) = \inf_{\mu} J^{\mu}(x_0) = J^{\mu^*}(x_0), \tag{1.5}$$

where μ^* denotes the optimal policy.

Dynamic programming is based on Bellman's principle of optimality [1, 13, 24]: An optimal (control) policy has the property that no matter what previous decisions have been, the remaining decisions must constitute an optimal policy with regard to the state resulting from those previous decisions.

According to Bellman, the minimum cost of any state starting at time k consists of two parts, one of which is the minimum cost at time k and the other part is the cumulative sum of the infinite minimum cost starting from time $k + 1$. In terms of equations, this means that

$$\begin{aligned} J^*(x_k) &= \min_{u_k}\{U(x_k, u_k) + \gamma J^*(x_{k+1})\} \\ &= \min_{u_k}\{U(x_k, u_k) + \gamma J^*(F(x_k, u_k))\}. \end{aligned} \tag{1.6}$$

This is known as the Bellman optimality equation, or the discrete-time Hamilton-Jacobi-Bellman (HJB) equation. One then has the optimal policy, i.e., the optimal control u_k^* at time k is the u_k that achieves this minimum as

$$u^* = \arg\min_{u_k} J\{U(x_k, u_k) + \gamma J^*(x_{k+1})\}. \tag{1.7}$$

Since one must know the optimal policy at time $k + 1$ to (1.6) use to determine the optimal policy at time k, Bellman's principle yields a backwards-in-time procedure for solving the optimal control problem. It is the basis for dynamic programming algorithms in extensive use in control system theory, operations research, and elsewhere.

1.1.2 ADP for Continuous-Time Systems

For continuous-time systems, the cost function J is also the key to dynamic programming. By minimizing J, one gets the optimal cost function J^*, which is often a Lyapunov function of the system. As a consequence of the Bellman's principle of optimality, J^*satisfies the Hamilton-Jacobi-Bellman (HJB) equation. But usually, one cannot get the analytical solution of the HJB equation. Even to find an accurate numerical solution is very difficult due to the so-called curse of dimensionality.

Consider the continuous-time nonlinear dynamical system

$$\dot{x}(t) = F(x(t), u(t)), t \geq t_0, \tag{1.8}$$

where $x \in \mathbb{R}^n$ is the state vector and $u \in \mathbb{R}^m$ is the control input vector. The corresponding cost function of the system can be defined as

$$J(x_0, u) = \int_{t_0}^{\infty} U(x(\tau), u(\tau)) d\tau, \tag{1.9}$$

with utility function $U(x, u) \geq 0$, where $x(t_0) = x_0$. The Bellman's principle of optimality can also be applied to continuous-time systems. In this case, the optimal cost

$$J^*(x(t)) = \min_{u(t)}\{J(x(t), u(t))\}, t \geq t_0, \tag{1.10}$$

satisfies the HJB equation

$$-\frac{\partial J^*}{\partial t} = \min_{u(t)}\left\{U(x, u) + \left(\frac{\partial J^*}{\partial x}\right)^T F(x, u)\right\}. \tag{1.11}$$

The HJB equation in (1.11) can be derived from the Bellman's principle of optimality [24]. Meanwhile, the optimal control $u^*(t)$ will be the one that minimizes the cost function,

$$u^*(t) = \arg\min_{u(t)}\{J(x(t), u(t))\}, t \geq t_0. \tag{1.12}$$

In 1994, Saridis and Wang [25] studied the nonlinear stochastic systems described by

$$\mathrm{d}x = f(x, t)\mathrm{d}t + g(x, t)u\,\mathrm{d}t + h(x, t)\mathrm{d}w, t_0 \leq t \leq T \tag{1.13}$$

with the cost function

$$J(x_0, u) = \mathbb{E}\left\{\int_{t_0}^{T} \left(Q(x, t) + u^T u\right) \mathrm{d}t + \phi(x(T), T) : x(t_0) = x_0\right\}$$

where $x \in \mathbb{R}^n, u \in \mathbb{R}^m$, and $w \in \mathbb{R}^k$ are the state vector, the control vector, and a separable Wiener process; f, g and h are measurable system functions; and Q and ϕ are nonnegative functions. A value function V is defined as

$$V(x,t) = \mathbb{E}\left\{\int_t^T \left(Q(x,t) + u^T u\right) \mathrm{d}t + \phi(x(T),T) : x(t_0) = x_0\right\}, t \in I,$$

where $I \triangleq [t_0, T]$. The HJB equation is modified to become the following equation

$$\frac{\partial V}{\partial t} + \mathscr{L}_u V + Q(x,t) + u^T u = \nabla V \tag{1.14}$$

where $\mathscr{L}_u$ is the infinitesimal generator of the stochastic process specified by (1.13) and is defined by

$$\begin{aligned}\mathscr{L}_u V =& \frac{1}{2}\operatorname{tr}\left\{h(x,t)h^T(x,t)\frac{\partial}{\partial x}\left(\frac{\partial V(x,t)}{\partial x}\right)^T\right\} \\ &+ \left(\frac{\partial V(x,t)}{\partial x}\right)^T (f(x,t) + g(x,t)u)\end{aligned}$$

Depending on whether $\nabla V \leq 0$ or $\nabla V \geq 0$, an upper bound $\bar{V}$ or a lower bound $\underline{V}$ of the optimal cost J^* are found by solving equation (1.14) such that $\underline{V} \leq J^* \leq \bar{V}$. Using $\bar{V}$ (or $\underline{V}$) as an approximation to J^*, one can solve for a control law. This leads to the so-called suboptimal control. It was proved that such controls are stable for the infinite-time stochastic regulator optimal control problem, where the cost function is defined as

$$J(x_0, u) = \lim_{T\to\infty} \mathbb{E}\left\{\frac{1}{T}\int_{t_0}^T \left(Q(x,t) + u^T u\right) \mathrm{d}t : x(t_0) = x_0\right\}$$

The benefit of the suboptimal control is that the bound V of the optimal cost J^* can be approximated by an iterative process. Beginning from certain chosen functions u_0 and V_0, let

$$u_i(x,t) = -\frac{1}{2}g^T(x,t)\frac{\partial V_{i-1}(x,t)}{\partial x}, i = 1, 2, \ldots. \tag{1.15}$$

Then, by repeatedly applying (1.14) and (1.15), one will get a sequence of functions V_i. This sequence $\{V_i\}$ will converge to the bound $\bar{V}$ (or V) of the cost function J^*. Consequently, u_i will approximate the optimal control when i tends to ∞. It is important to note that the sequences $\{V_i\}$ and $\{u_i\}$ are obtainable by computation and they approximate the optimal cost and the optimal control law, respectively.

Some further theoretical results for ADP have been obtained in [2]. These works investigated the stability and optimality for some special cases of ADP. In [2], Murray et al. studied the (deterministic) continuous-time affine nonlinear systems

$$\dot{x} = f(x) + g(x)u, x(t_0) = x_0 \tag{1.16}$$

with the cost function

$$J(x, u) = \int_{t_0}^{\infty} U(x, u)\mathrm{d}t \tag{1.17}$$

where $U(x, u) = Q(x) + u^T R(x)u$, $Q(x) > 0$ for $x \neq 0$ and $Q(0) = 0$, and $R(x) > 0$ for all x. Similar to [25], an iterative procedure is proposed to find the control law as follows. For the plant (1.16) and the cost function (1.17), the HJB equation leads to the following optimal control law

$$u^*(x) = -\frac{1}{2}R^{-1}(x)g^T(x)\left[\frac{\mathrm{d}J^*(x)}{\mathrm{d}x}\right]. \tag{1.18}$$

Applying (1.17) and (1.18) repeatedly, one will get sequences of estimations of the optimal cost function J^* and the optimal control u^*. Starting from an initial stabilizing control $v_0(x)$, for $i = 0, 1, \ldots$, the approximation is given by the following iterations between value functions

$$V_{i+1}(x) = \int_t^{\infty} U\left(x(\tau), v_i(\tau)\right) \mathrm{d}\tau$$

and control laws

$$v_{i+1}(x) = -\frac{1}{2}R^{-1}(x)g^T(x)\left[\frac{\mathrm{d}V_{i+1}(x)}{\mathrm{d}x}\right]$$

The following results were shown in [2].

(1) The sequence of functions $\{V_i\}$ obtained above converges to the optimal cost function J^*.

(2) Each of the control laws v_{i+1} obtained above stabilizes the plant (1.16), for all $i = 0, 1, \ldots$

(3) Each of the value functions $V_{i+1}(x)$ is a Lyapunov function of the plant, for all $i = 0, 1, \ldots$

Abu-Khalaf and Lewis [26] also studied the system (1.16) with the following value function

$$V(x(t)) = \int_t^{\infty} U(x(\tau), u(\tau))\mathrm{d}\tau = \int_t^{\infty} \left(x^T(\tau)Qx(\tau) + u^T(\tau)Rx(\tau)\right)\mathrm{d}\tau$$

where Q and R are positive-definite matrices. The successive approximation to the HJB equation starts with an initial stabilizing control law $v_0(x)$. For $i = 0, 1, \ldots$, the approximation is given by the following iterations between policy evaluation

$$0 = x^T Qx + v_i^T(x)Rv_i(x) + \nabla V_i^T(x)\left(f(x) + g(x)v_i(x)\right)$$

and policy improvement

$$v_{i+1}(x) = -\frac{1}{2}R^{-1}g^T(x)\nabla V_i(x)$$

where $\nabla V_i(x) = \partial V_i(x)/\partial x$. In [26], the above iterative approach was applied to systems (1.16) with saturating actuators through a modified utility function, with convergence and optimality proofs showing that $V_i \to J^*$ and $v_i \to u^*$, as $i \to \infty$. For continuous-time optimal control problems, attempts have been going on for a long time in the quest for successive solutions to the HJB equation. Published works can date back to as early as 1967 by Leake and Liu [26]. The brief overview presented here only serves as a beginning of many more recent results [26–28].

1.2 Publication Outline

The general layout of the presentation of this monograph is given as follows. Adaptive dynamic programming is used to design drug dosage regulation mechanisms to provide adaptive viral treatment strategies for input-limited organisms, and to extend this to tumour cells, immune cells and interplay and regulation schemes among the immune system. The main contents of this monograph are shown as follows:

Chapter 1 introduces the research background, development and current status of ADP both domestically and internationally, as well as the idea and design framework of the underlying ADP, including discrete-time and continuous-time systems.

Chapter 2 investigates optimal regulation scheme between tumor and immune cells based on ADP approach. The therapeutic goal is to inhibit the growth of tumor cells to allowable injury degree, and maximize the number of immune cells in the meantime. The reliable controller is derived through the ADP approach to make the number of cells achieve the specific ideal states. Firstly, the main objective is to weaken the negative effect caused by chemotherapy and immunotherapy, which means that minimal dose of chemotherapeutic and immunotherapeutic drugs can be operational in the treatment process. Secondly, according to nonlinear dynamical mathematical model of tumor cells, chemotherapy and immunotherapeutic drugs can act as powerful regulatory measures, which is a closed-loop control behavior. Finally, states of the system and critic weight errors are proved to be ultimately uniformly bounded with the appropriate optimization control strategy and the simulation results are shown to demonstrate effectiveness of the cybernetics methodology.

Chapter 3 investigates the optimal control strategy problem for nonzero-sum games of the immune system based on adaptive dynamic programming. Firstly, the main objective is approximating a Nash equilibrium between the tumor cells and the immune cell population, which is governed through chemotherapy drugs and immunoagents guided by the mathematical growth model of the tumor cells. Secondly, a novel intelligent nonzero-sum games-based ADP is put forward to solve optimization control problem through reducing the growth rate of tumor cells and minimizing chemotherapy drugs and immunotherapy drugs. Meanwhile, convergence analysis and iterative ADP algorithm are specified to prove feasibility. Finally, simulation examples are listed to account for availability and effectiveness of the research methodology.

Chapter 4 devotes to evolutionary dynamics optimal control oriented tumor immune differential game system. Firstly, the mathematical model covering immune cells and tumor cells considering the effects of chemotherapy drugs and immune agents. Secondly, the bounded optimal control problem covering is transformed into solving HJB equation considering the actual constraints and infinite-horizon performance index based on minimize the amount of medication administered. Finally, approximate optimal control strategy is acquired through iteration dual heuristic dynamic programming algorithm avoiding dimensional disaster effectively and providing optimal treatment scheme for clinical applications.

Chapter 5 mainly proposes an evolutionary algorithm and its first application to develop therapeutic strategies for Ecological Evolutionary Dynamics Systems (EEDS), obtaining the balance between tumor cells and immune cells by rationally arranging chemotherapeutic drugs and immune drugs. Firstly, an EEDS nonlinear kinetic model is constructed to describe the relationship between tumor cells, immune cells, dose, and drug concentration. Secondly, the N-Level Hierarchy Optimization (NLHO) algorithm is designed and compared with 5 algorithms on 20 benchmark functions, which proves the feasibility and effectiveness of NLHO. Finally, we apply NLHO into EEDS to give a dynamic adaptive optimal control policy, and develop therapeutic strategies to reduce tumor cells, while minimizing the harm of chemotherapy drugs and immune drugs to the human body. The experimental results prove the validity of the research method.

Chapter 6 investigates the optimal control strategy for organism by using ADP method under the architecture of Firstly, a tumor model is established to formulate the interaction relationships among normal cells, tumor cells, endothelial cells and the concentrations of drugs. Then, the ADP-based method of single-critic network architecture is proposed to approximate the coupled HJEs under the medicine dosage regulation mechanism (MDRM). According to game theory, the approximate MDRM-based optimal strategy can be derived, which is of great practical significance. Owing to the proposed mechanism, the dosages of the chemotherapy and anti-angiogenic drugs can be regulated timely and necessarily. Furthermore, the stability of the closed-loop system with the obtained strategy is analyzed via Lyapunov theory. Finally, a simulation experiment is conducted to verify the effectiveness of the proposed method.

Chapter 7 investigates the constrained adaptive control strategy based on virotherapy for organism using the MDRM. Firstly, the tumor-virus-immune interaction dynamics is established to model the relations among the tumor cells (TCs), virus particles and the immune response. ADP method is extended to approximately obtain the optimal strategy for the interaction system to reduce the populations of TCs. Due to the consideration of asymmetric control constraints, the non-quadratic functions are proposed to formulate the value function such that the corresponding Hamilton-Jacobi-Bellman equation (HJBE) is derived which can be deemed as the cornerstone of ADP algorithms. Then, the ADP method of single-critic network architecture which integrates MDRM is proposed to obtain the approximate solutions of HJBE and eventually derive the optimal strategy. The design of MDRM makes it possible for the dosage of the agentia containing oncolytic virus particles to be regulated timely and necessarily. Furthermore, the uniform ultimate boundedness of the system states and critic weight estimation errors are validated by Lyapunov stability analysis. Finally, simulation results are given to show the effectiveness of the derived therapeutic strategy.

References

1. Bellman RE (2010) Dynamic programming. Princeton University Press
2. Murray JJ, Cox CJ, Lendaris GG, Saeks R (2002) Adaptive dynamic programming. IEEE Trans Syst Man Cybernet Part C (Appl Rev) 32(2):140–153
3. Lewis FL, Vrabie D (2009) Reinforcement learning and adaptive dynamic programming for feedback control. IEEE Circuits Syst Mag 9(3):32–50
4. Zhang HG, Zhang X, Luo YH, Yang J (2013) An overview of research on adaptive dynamic programming. Acta Automatica Sinica 39(4):303–311
5. Jiang ZP, Jiang Y (2013) Robust adaptive dynamic programming for linear and nonlinear systems: an overview. Eur J Control 19(5):417–425
6. He H, Ni Z, Fu J (2012) A three-network architecture for on-line learning and optimization based on adaptive dynamic programming. Neurocomputing 78(1):3–13
7. Bertsekas DP (2015) Value and policy iterations in optimal control and adaptive dynamic programming. IEEE Trans Neural Netw Learn Syst 28(3):500–509
8. Luo B, Liu D, Wu HN, Wang D, Lewis FL (2017) Policy gradient adaptive dynamic programming for data-based optimal control. IEEE Trans Cybernet 47(10):3341–3354
9. Yang Y, Vamvoudakis KG, Modares H, Yin Y, Wunsch DC (2021) Hamiltonian-driven hybrid adaptive dynamic programming. IEEE Trans Syst Man, Cybernet: Syst 51(10):6423–6434
10. Jiang Y, Jiang ZP (2013) Robust adaptive dynamic programming with an application to power systems. IEEE Trans Neural Netw Learn Syst 24(7):1150–1156
11. Jiang Y, Jiang ZP (2015) Global adaptive dynamic programming for continuous-time nonlinear systems. IEEE Trans Autom Control 60(11):2917–2929
12. Bertsekas DP, Tsitsiklis JN (1996) Neuro-dynamic programming. Athena Scientific
13. Bertsekas DP (2011) Dynamic programming and optimal control, vol ii, 3rd edn. Athena Scientific, Belmont, MA
14. Si J, Barto AG, Powell WB, Wunsch D (2004) Handbook of learning and approximate dynamic programming. Wiley
15. Powell WB (2007) Approximate dynamic programming: solving the curses of dimensionality. Wiley
16. Balakrishnan SN, Ding J, Lewis FL (2008) Issues on stability of ADP feedback controllers for dynamical systems. IEEE Trans Syst Man Cybernet Part B: Cybernet 38(4):913–917
17. Wang FY, Zhang HG, Liu DR (2009) Adaptive dynamic programming: an introduction. IEEE Comput Intell Mag 4(2):39–47

18. Duan J, Liu Z, Li SE, Sun Q, Jia Z, Cheng B (2022) Adaptive dynamic programming for nonaffine nonlinear optimal control problem with state constraints. Neurocomputing 484:128–141
19. Zhao J, Na J, Gao G (2022) Robust tracking control of uncertain nonlinear systems with adaptive dynamic programming. Neurocomputing 471:21–30
20. Jiang H, Zhou B (2022) Bias-policy iteration based adaptive dynamic programming for unknown continuous-time linear systems. Automatica 136:110058
21. Ming Z, Zhang H, Li W, Luo Y (2022) Neurodynamic programming and tracking control for nonlinear stochastic systems by PI algorithm. IEEE Trans Circuits Syst II Express Briefs 69(6):2892–2896
22. Dou L, Cai S, Zhang X, Su X, Zhang R (2022) Event-triggered-based adaptive dynamic programming for distributed formation control of multi-UAV. J Frankl Inst 359(8):3671–3691
23. Wang R, Ma D, Li MJ, Sun Q, Zhang H, Wang P (2022) Accurate current sharing and voltage regulation in hybrid wind/solar systems: an adaptive dynamic programming approach. IEEE Trans Consum Electron 68(3):261–272
24. Lewis FL, Syrmos VL (1995) Optimal control. Wiley, New Yorks
25. Saridis GN, Wang FY (1994) Suboptimal control of nonlinear stochastic systems. Control Theory Adv Technol 10(4):847–871
26. Abu-Khalaf M, Lewis FL (2005) Nearly optimal control laws for nonlinear systems with saturating actuators using a neural network HJB approach. Automatica 41(5):779–791
27. Yang X, Liu D, Huang Y (2013) Neural-network-based online optimal control for uncertain non-linear continuous-time systems with control constraints. IET Control Theory Appl 7(17):2037–2047
28. Yang X, Liu D, Wang D (2014) Reinforcement learning for adaptive optimal control of unknown continuous-time nonlinear systems with input constraints. Int J Control 87(3):553–556

Chapter 2
Neural Networks-Based Immune Optimization Regulation Using Adaptive Dynamic Programming

2.1 Introduction

In the fight against cancer, there had been no effective measures before chemotherapy and radiation appeared since there only exist tiny differences between cancer cells and normal cells. Doctors operate to remove solid tumors that have not yet spread, which can not guarantee cancer from recurring. When radiotherapy and chemotherapy have increased side effects, and targeted therapy is not flexible because of its strong pertinence, the scientific research direction began to turn to the human body system. Generally, tumor cells escape from the immune system, not because it fails for the immune system to recognize them or it is not activated, but cancer cells have evolved a way to block the activation of T cells by making a specific binding. Thus, the medical communities have struggled to find a lot of special means for cancer cells to intercept the activation of the T cells, freeing up the immune system. Compared with traditional treatments such as surgery, radiation and chemotherapy, immunotherapy has fewer side effects and better therapeutic effects. However, it is difficult to tackle the transient period of immune agents. Therefore, the hybrid therapy of chemotherapy and immunotherapy is a better choice. As [1], it is hardly sufficient to control tumor growth through neither chemotherapy nor immunotherapy alone, but tumor cells can be eradicated by adopting the combination therapies which is known as biochemotherapy described in [2].

With extensive development of nonlinear dynamic [3, 4], its engineering application scenarios enjoy increasing diversification such as competitive Nash equilibrium problems, especially in the biomedical field. And not coincidentally, game theory has been introduced into the interaction model of tumor cells and immune cells. Both of the chemotherapy and immunotherapy aim at reducing the number of tumor cells. Based on this fact, the collaborative game is formed and one can design adaptive therapy from the view of game theory. Multiple biological interactions constitute complex nonlinear growth process of tumor cells, however, regarding major influence factors of tumor cell populations as research object is the focus. Hunting cells refer to the immune cells participating in removing foreign agents and strengthening

J. Sun et al., *Adaptive Dynamic Programming*,
https://doi.org/10.1007/978-981-99-5929-7_2

the immune response. temperatures have suggested that cell-mediated anti-tumor immunity contributed to increasing the population of hunting tumor cells to maintain a specific proportion between the resting and the hunting predator cells as 40% in literature [5], which was beneficial for maintenance of the tumor dormant state. The immune regulations vary from individual to individual, but immunotherapy-based optimal regulation plays the role of reducing tumor cells without considering certain circumstances in case of special invocation. Enhanced tumor antigen presentation could effectively stimulate dendritic cells and increase the immunotherapy-based curative effect in [6]. The known "predator-prey" between immune cells and tumor cells leads to cyclic growth and reduction, which can be continue indefinitely or reach an equilibrium saddle point determined by system parameters. Literature [7] investigated nonlinear dynamical model which provided guiding significance for introducing that into cybernetics. As known, system identification or optimal control is of great practical value. As a powerful and effective optimization algorithm, the ADP method can solve the nonlinear optimal control problems well, realizing the most appropriate therapeutic strategy.

Of course, the immune system has the responsibility for restraining tumor growth, but it is hardly to fight out the tumor cells alone. Firstly, ego characteristic of tumor cells compared to normal cells within the body leads to no exclusion and tolerance to tumor cells of the immune system. Secondly, there is no strong defense mechanism itself in fighting with the cancer cells which means the failure of the immune response. Finally, Immune function was observed to be protective through intervention with organic binding agents of CD4 and CD8 cells. Chemotherapy can not only rapidly kill differentiated tumor cells, but also destroy regular cells. This side effect caused by chemotherapy can be lessened through introducing the immunotherapy. Thus the combined therapy of chemotherapy and immunotherapy is more reasonable. Immunotherapies can strengthen the immune system through extra stimulation, on the other hand, improve the ability to recognize foreign entity. Therefore, decelerating the growth rate of tumor cells with minimized dose of chemotherapy and immunotherapeutic drugs is the control objective. Furthermore, optimal control strategy is obtained through ADP method, giving the optimal levels of each treatment regimen through nonzero-sum differential games strategy developed in [8].

Prescribed performance tracking control has been creatively developed as [9], however, there is seldom any literatures focusing on this scope considers mutual relationship among tumor cells, immune cells, chemotherapy and immunotherapy drugs, let alone setting the performance as eventually acquired of optimal therapeutic effect associated with coupling behaviors mentioned above. Retrospect to literatures as [10], the chapter transformed it into multi-player nonzero-sum games problems whose optimal control was obtained by complex decoupling in dealing with Hamilton-Jacobi equation as [11]. Subsequently, online adaptive and off-policy learning algorithms were respectively developed in [12–14]. Of course, the constrained-input was taken into consideration, when it comes to practical applications in [15], even more intensive work on uncertain constraints were in contemplation considered as [16]. As [17], the control policies of the distributed subsystems acted as players, noticeably, the chapter was formulated as a two-players nonzero-sum game including chemotherapy and immunotherapy. [18] first introduced an updating strategy based on intertask relationships. Synchronously, reciprocal action between the tumor cells and immune cells which could be analogous to interactions between systems in [19, 20].

The unknown nonlinear dynamic is usually implemented by fuzzy control as [21, 22] and neural networks in [18, 23], where the actor network and critic network are adopted for updating control policy at an appropriate time through policy iteration technique as [24–26]. The convergence of model-based policy iteration algorithm is equivalent to that of data-based learning as [27]. Similarly, states of the system and critic error are required to be ultimately uniformly bounded during the process of value iteration, which is guaranteed through event-triggered formation control scheme firstly proposed for all signals of the closed-loop system in literature [28]. According to the iterative value algorithm, the optimum can be obtained through learning continuously [29, 30]. However there is little research on the two-players nonzero-sum game considering tumor cells and immune cells using the proposed value iteration learning.

2.2 Preliminaries

As is known, there exist interaction relationships among the anticancer agent cells, lymphocytes and macrophages that constitute the basic immune system microenvironment, which can be presented as follows. Firstly, T-lymphocytes and cytotoxic macrophages/natural killer cells can effectively damage tumor cells. Secondly, destroyed behaviour of macrophages can also active T-lymphocytes for launching another attack. Meanwhile, the population of T-lymphocytes can be fed through resting cells. Finally, the model is guided by degradation of resting cells and activation of immune cells by natural growth rate. This section gives the nonlinear growth equation which can represent the whole immune response.

$$N_{total} = \frac{\upsilon N_H(t) N_T(t)}{\nu + N_T(t)} \tag{2.1}$$

where $N_H(t)$, $N_T(t)$ denote the number of hunting cells and tumor cells at time t, respectively. υ and ν are positive constants. The changes in quantity caused by the inactivation of the immune cells and the apoptosis of tumor cells are presented as:

$$\begin{aligned} \frac{dN_T(t)}{dt} &= -\sigma_1 N_H(t) N_T(t) \\ \frac{dN_H(t)}{dt} &= -\sigma_2 N_H(t) N_T(t) \end{aligned} \tag{2.2}$$

where σ_1 denotes the loss rate of $N_T(t)$ caused by $N_H(t)$ and σ_2 represents the loss rate of $N_H(t)$ caused by $N_T(t)$. The situations above reflect the competition between tumor cells and the host cells. Then we construct the dynamic equations as follows

Table 2.1 Detailed descriptions of system parameters

Parameter	Estimated value
ι_1	Intrinsic growth rate of $N_T(t)$ ignoring $N_{CD}(t)$
ϱ_1	Reciprocal carrying capacity of $N_T(t)$ irrespective of $N_T(t)$ and $N_{CD}(t)$
$\hbar$	Constant influx rate of $N_H(t)$
σ_1	Rate of loss of $N_T(t)$ for $N_H(t)$
δ_1	Response coefficient to $N_{CD}(t)$ for $N_T(t)$
$\mathfrak{D}$	Per capita decay rate of $N_H(t)$ without regard to $N_T(t)$, $N_{CD}(t)$ and $N_{ID}(t)$
σ_2	Rate of loss of $N_H(t)$ for $N_T(t)$
δ_2	Response coefficient to $N_{CD}(t)$ for $N_H(t)$
υ	Maximum recruitment rate of $N_H(t)$ by ligand-transduced $N_T(t)$
ς	Maximum recruitment rate of $N_H(t)$ by $N_{ID}(t)$
ν	Steepness coefficient of $N_H(t)$ by $N_T(t)$
ϑ	Steepness coefficient of $N_H(t)$ by $N_{ID}(t)$
φ_1	Decay rate of $N_{CD}(t)$
φ_2	Decay rate of $N_{ID}(t)$

$$\begin{aligned}\dot{N}_T(t) =& \iota_1 N_T(t)(1-\varrho_1 N_T(t)) - \sigma_1 N_T(t) N_H(t)\\ &- \delta_1 N_{CD}(t) N_T(t)\\ \dot{N}_H(t) =& \frac{\upsilon N_H(t) N_T^2(t)}{\nu + N_T^2(t)} + \frac{\varsigma N_H(t) N_{ID}(t)}{\vartheta + N_{ID}(t)} - \sigma_2 N_T(t) N_H(t)\\ &- \mathfrak{D} N_H(t) - \delta_2 N_{CD}(t) N_H(t)\end{aligned} \tag{2.3}$$

where $\mathfrak{D}$ represents the death rate of cells without considering any tumor cells. ι_α ($\alpha = 1, 2$) and ϱ_α denote the per capita growth rates and reciprocal carrying capacities. The descriptions of the other associated parameters are given in Table 2.1.

Consider the given chemotherapy and immunotherapy drugs as $u(t)$ and $v(t)$ at time t, which is regarded as multiple dose administration compared with influence of recombinant human interleukin-11 for injection or recombinant human granulocyte colony-stimulating factor injection. Assume that targeted therapy cannot be achieved through only chemotherapeutic drugs. Then we can obtain that

$$f_{response}(t) = s_\alpha(1 - e^{-\lambda u(t)}) \tag{2.4}$$

where s_α is the different response coefficients for distinguishing the change rate of different cells. The mathematical model considering injected drugs is presented as

$$
\begin{aligned}
\dot{N}_{CD}(t) =& u(t) - \varphi_1 N_{CD}(t) \\
\dot{N}_{ID}(t) =& v(t) - \varphi_2 N_{ID}(t) \\
\dot{N}_T(t) =& \iota_1 N_T(t)(1 - \varrho_1 N_T(t)) - \sigma_1 N_T(t) N_H(t) \\
& - \delta_1 N_{CD}(t) N_T(t) - s_2(1 - e^{-\lambda u(t)}) \\
\dot{N}_H(t) =& \frac{\upsilon N_H(t) N_T^2(t)}{\nu + N_T^2(t)} + \frac{\varsigma N_H(t) N_{ID}(t)}{\vartheta + N_{ID}(t)} - \sigma_2 N_T(t) N_H(t) - \mathfrak{D} N_H(t) \\
& - \delta_2 N_{CD}(t) N_H(t) - s_1(1 - e^{-\lambda u(t)})
\end{aligned}
\tag{2.5}
$$

where $N_{CD}(t)$ and $N_{ID}(t)$ are concentrations of chemotherapy and immunotherapy. $v(t)$ and $u(t)$ are the doses of chemotherapeutic drug and immunotherapeutic drug. Generally speaking, λ is taken as 1 for the unknown role of cytokines.

Remark 2.1 The model (2.5) describes the relations among the hunting cells, the tumor cells, the concentration of chemotherapy agentia, and the concentration of immunotherapy agentia. From (2.5) we can find both of the hunting cells and the chemotherapy agentia can reduce the number of tumor cells, and the immunotherapy agentia can stimulate the growth of hunting cells. On the other hand, the tumor cells can influence the number of hunting cells. Based on this complicated interactive relationship, we can obtain the optimal object through ADP, that is, minimization of tumor cells while ensuring the number of normal cells at certain time t.

Before proceeding, let $X = [N_T, N_H, N_{CD}, N_{ID}]^T$, then the model (2.5) can be simplified as

$$
\dot{X}(t) = f(X) + g(X)u(t) + \kappa(X)v(t) \tag{2.6}
$$

where $f(X)$ is the right-hand dynamics of (2.5) excluding the control $u(t)$ and $v(t)$. The matrixes $g(X) = [0, 0, 1, 0]^T$ and $\kappa(X) = [0, 0, 0, 1]^T$.

For system (2.6), the performance index function of the ϵ player can be given as

$$
J_\epsilon(X_0) = \int_0^\infty \Big(X^T \mathcal{Q}_\epsilon X + u^T \mathcal{R}_{\epsilon 1} u + v^T \mathcal{R}_{\epsilon 2} v\Big) d\tau \tag{2.7}
$$

where $\mathcal{Q}_\epsilon$ is positive definite matrix, $\mathcal{R}_{\epsilon 1}$ and $\mathcal{R}_{\epsilon 2}$ are symmetric positive matrixes. The corresponding cost functions are presented as:

$$
\mathcal{V}_\epsilon(X, u, v) = \int_t^\infty \mathfrak{R}_\epsilon(X, u, v) d\tau \tag{2.8}
$$

with the utility function

$$
\mathfrak{R}_\epsilon(X, u, v) = X^T \mathcal{Q}_\epsilon X + u^T \mathcal{R}_{\epsilon 1} u + v^T \mathcal{R}_{\epsilon 2} v. \tag{2.9}
$$

Definition 2.2 For two-player NZS game of system (2.6), the Nash equilibrium solution is said to be obtained with the control pair (u^*, v^*) which satisfied that,

$$\begin{aligned} \mathcal{V}_\epsilon(u^*, v^*) &\le \mathcal{V}_\epsilon(u, v^*) \\ \mathcal{V}_\epsilon(u^*, v^*) &\le \mathcal{V}_\epsilon(u^*, v) \end{aligned} \tag{2.10}$$

for any admissible control policies u and v.

The Hamilton functions can be constructed as:

$$\begin{aligned} \mathrm{H}_\epsilon(X, u, v) = {} & X^T \mathcal{Q}_\epsilon X + u^T \mathcal{R}_{\epsilon 1} u + v^T \mathcal{R}_{\epsilon 2} v \\ & + \nabla \mathcal{V}_\epsilon^T (f(X) + \boldsymbol{g}(X)u(t) + \kappa(X)v(t)) \end{aligned} \tag{2.11}$$

where $\nabla \mathcal{V}_\epsilon$ is the partial derivative of the cost function and $\epsilon = 1, 2$. According to the stationarity conditions at equilibrium points, the optimal control for two players are obtained

$$\begin{aligned} u^* &= -\frac{1}{2} \mathcal{R}_{11}^{-1} \boldsymbol{g}^T(X) \nabla \mathcal{V}_1^* \\ v^* &= -\frac{1}{2} \mathcal{R}_{22}^{-1} \kappa^T(X) \nabla \mathcal{V}_2^* \end{aligned} \tag{2.12}$$

with $\mathcal{V}_1^*$ and $\mathcal{V}_2^*$ being the solutions of coupled HJ equations as

$$\begin{aligned} X^T \mathcal{Q}_1 X & - \frac{1}{4} \nabla \mathcal{V}_1^{*T} \boldsymbol{g}(X) \mathcal{R}_{11}^{-1} \boldsymbol{g}^T(X) \nabla \mathcal{V}_1^* + \nabla \mathcal{V}_1^{*T} f(X) \\ & + \frac{1}{4} \nabla \mathcal{V}_2^{*T} \kappa(X) \mathcal{R}_{22}^{-1} \mathcal{R}_{12} \mathcal{R}_{22}^{-1} \kappa^T(X) \nabla \mathcal{V}_2^* \\ & - \frac{1}{2} \nabla \mathcal{V}_1^{*T} \kappa(X) \mathcal{R}_{22}^{-1} \kappa^T(X) \nabla \mathcal{V}_2^* = 0, \end{aligned} \tag{2.13}$$

and

$$\begin{aligned} X^T \mathcal{Q}_2 X & - \frac{1}{4} \nabla \mathcal{V}_2^{*T} \kappa(X) \mathcal{R}_{22}^{-1} \kappa^T(X) \nabla \mathcal{V}_2^* + \nabla \mathcal{V}_2^{*T} f(X) \\ & + \frac{1}{4} \nabla \mathcal{V}_1^{*T} \boldsymbol{g}(X) \mathcal{R}_{11}^{-1} \mathcal{R}_{21} \mathcal{R}_{11}^{-1} \boldsymbol{g}^T(X) \nabla \mathcal{V}_1^* \\ & - \frac{1}{2} \nabla \mathcal{V}_2^{*T} \boldsymbol{g}(X) \mathcal{R}_{11}^{-1} \boldsymbol{g}^T(X) \nabla \mathcal{V}_1^* = 0. \end{aligned} \tag{2.14}$$

Lemma 2.3 *For nonlinear system (2.6), suppose that $\mathcal{V}_1^*$ and $\mathcal{V}_2^*$ satisfy the equations (2.13) and (2.14). Then under the optimal control (2.12), the system is asymptotically stable.*

Proof The proof is omitted since it is similar to that in [31, 32].

By solving the coupled HJ equations (2.13) and (2.14), one can obtain the optimal control as (2.12), which means the Nash equilibrium for the two-player NZS game system is attained. Nevertheless, due to the existence of nonlinear terms and coupled terms, these partial differential equations are uneasy to solve. Since ADP is a powerful approximate learning method, the approximate solutions of (2.13) and (2.14) can be acquired.

2.3 Design of Adaptive Controller

In order to find the optimal control strategy, a critic network is constructed based on neural network firstly. And then optimal value function can be shown as:

$$\mathcal{V}_\epsilon^* = (\zeta_\epsilon^*)^T \xi_\epsilon(X) + o_\epsilon, \epsilon = 1, 2, \tag{2.15}$$

where $\zeta_\epsilon^* \in R^{p_\epsilon}$, $\xi_\epsilon \in R^{p_\epsilon}$ and $o_\epsilon \in$ R are the ideal weight vector, activation function and approximation error of the neural network. As it's scarcely possible to get the weight ζ_ϵ^*, we give the approximate version

$$\hat{\mathcal{V}}_\epsilon^* = (\hat{\zeta}_\epsilon)^T \xi_\epsilon(X). \tag{2.16}$$

Based on (2.12) and (2.15), we obtain the optimal control as

$$\begin{aligned} u^* &= -\frac{1}{2}\mathcal{R}_{11}^{-1} \boldsymbol{g}^T(X)((\nabla\xi_1(X))^T \zeta_1^* + \nabla o_1) \\ v^* &= -\frac{1}{2}\mathcal{R}_{22}^{-1} \kappa^T(X)((\nabla\xi_2(X))^T \zeta_2^* + \nabla o_2) \end{aligned} \tag{2.17}$$

Then we further get the approximate control policies as

$$\begin{aligned} \hat{u} &= -\frac{1}{2}\mathcal{R}_{11}^{-1} \boldsymbol{g}^T(X)(\nabla\xi_1(X)^T \hat{\zeta}_1 \\ \hat{v} &= -\frac{1}{2}\mathcal{R}_{22}^{-1} \kappa^T(X)(\nabla\xi_2(X)^T \hat{\zeta}_2 \end{aligned} \tag{2.18}$$

Remark 2.4 For the unknowable nature of ideal weights, the NNs are used to approximate the system dynamics and approximate version as (2.16), aming at minimizing the current estimate of the value functions in (2.15) by selecting policies (2.18) can be obtained with available closed-form expressions.

According to (2.18), the closed-loop system can be rewritten as

$$\dot{X}(t) = f(X) + \boldsymbol{g}(X)\hat{u} + \kappa(X)\hat{v}. \tag{2.19}$$

Furthermore, we can attain the approximate Hamilton as

$$\begin{aligned} \mathrm{H}_\epsilon(X, \hat{u}, \hat{v}) &= X^T \mathcal{Q}_\epsilon X + \hat{u}^T \mathcal{R}_{\epsilon 1} \hat{u} \\ &\quad + \hat{v}^T \mathcal{R}_{\epsilon 2} \hat{v} + (\hat{\zeta}_\epsilon)^T \nabla \xi_\epsilon(X) \dot{X}(t) \\ &= e_\epsilon(t). \end{aligned} \tag{2.20}$$

To approach the optimal strategy and minimize $e_\epsilon(t)$, the goal of adaptive learning is set to be $\mathcal{E} = \mathcal{E}_1 + \mathcal{E}_2 = 1/2e_1^2 + 1/2e_2^2$. Then applying the gradient descent method, we obtain the learning law of critic for player ϵ

$$\dot{\hat{\zeta}}_\epsilon = -\varrho_\epsilon \frac{1}{(\delta_\epsilon^T \delta_\epsilon + 1)^2} \frac{\partial \mathcal{E}(t)}{\partial \hat{\zeta}_\epsilon} = -\varrho_\epsilon \frac{1}{(\delta_\epsilon^T \delta_\epsilon + 1)^2} \frac{\partial \mathcal{E}_\epsilon(t)}{\partial \hat{\zeta}_\epsilon} = -\varrho_\epsilon \frac{\delta_\epsilon e_\epsilon(t)}{(\delta_\epsilon^T \delta_\epsilon + 1)^2} \tag{2.21}$$

where $\delta_\epsilon = \nabla \xi_\epsilon(X) \dot{X}(t)$, and ϱ_ϵ is the positive learning law. Let $\tilde{\zeta}_\epsilon = \zeta_\epsilon^* - \hat{\zeta}_\epsilon$, then we have

$$\dot{\tilde{\zeta}}_\epsilon = \varrho_\epsilon \frac{\delta_\epsilon \sigma_\epsilon(t)}{(\delta_\epsilon^T \delta_\epsilon + 1)^2} - \varrho_\epsilon \frac{\delta_\epsilon \delta_\epsilon^T \tilde{\zeta}_\epsilon}{(\delta_\epsilon^T \delta_\epsilon + 1)^2} = \varrho_\epsilon \underline{\delta}_\epsilon \sigma_\epsilon(t) - \varrho_\epsilon \bar{\delta}_\epsilon \bar{\delta}_\epsilon^T \tilde{\zeta}_\epsilon, \tag{2.22}$$

where $\underline{\delta}_\epsilon = \delta_\epsilon/(\delta_\epsilon^T \delta_\epsilon + 1)^2$, $\bar{\delta}_\epsilon = \delta_\epsilon/(\delta_\epsilon^T \delta_\epsilon + 1)$ and $\sigma_\epsilon(t) = -\nabla o_\epsilon^T(X)(f(X) + g(X)\hat{u} + \kappa(X)\hat{v})$ is the approximate residual error when employing critic neural network [33].

Before presenting the main results of this chapter, two regular assumptions are necessary [34–36].

Assumption 2.1 For $\epsilon = 1, 2$, the signal $\bar{\delta}_\epsilon$ is persistently excited such that the following inequality is satisfied

$$\varsigma_\epsilon I_{\nu_\epsilon \times \nu_\epsilon} \leq \int_t^{t+T} \bar{\delta}_\epsilon \bar{\delta}_\epsilon^T d\varepsilon, \tag{2.23}$$

where ν_ϵ denotes the neuro number of the ϵth critic network.

Assumption 2.2 For $\epsilon = 1, 2$, there exist positive constants $\xi_{\epsilon max}$, $o_{\epsilon max}$ and $\sigma_{\epsilon max}$ such that the following inequalities hold, that is, $\|\nabla \xi_\epsilon(X)\| \leq \xi_{\epsilon max}$, $\|\nabla o\| \leq o_{\epsilon max}$ and $\|\sigma_\epsilon\| \leq \sigma_{\epsilon max}$.

Applying the Lyapunov method, the stability in the sense of UUB is demonstrated to be guaranteed by the following theorem.

Theorem 2.5 *For system (2.6), when the weight updating laws of critic networks are given by (2.21), then the UUB properties of the weight estimation error $\tilde{\zeta}_\epsilon$ can be guaranteed by the obtained control policies (2.18).*

Proof Select the Lyapunov function as

$$\mathcal{L} = \frac{1}{2}\varrho_1^{-1}\tilde{\zeta}_1^T\tilde{\zeta}_1^T + \frac{1}{2}\varrho_2^{-1}\tilde{\zeta}_2^T\tilde{\zeta}_2^T. \tag{2.24}$$

Taking the time derivative of (2.24), then we obtain

$$\begin{aligned}\dot{\mathcal{L}} &= \varrho_1^{-1}\tilde{\zeta}_1^T\dot{\tilde{\zeta}}_1 + \varrho_2^{-1}\tilde{\zeta}_2^T\dot{\tilde{\zeta}}_2 \\ &= \tilde{\zeta}_1^T(\underline{\delta}_1\sigma_1(t) - \bar{\delta}_1\bar{\delta}_1^T\tilde{\zeta}_1) + \tilde{\zeta}_2^T(\underline{\delta}_2\sigma_2(t) - \bar{\delta}_2\bar{\delta}_2^T\tilde{\zeta}_2)\end{aligned} \tag{2.25}$$

According to Young's inequality, we have

$$\tilde{\zeta}_1^T\underline{\delta}_1\sigma_1(t) \le \tilde{\zeta}_1^T\bar{\delta}_1\sigma_1(t) \le \frac{1}{2}\tilde{\zeta}_1^T\bar{\delta}_1\bar{\delta}_1^T\tilde{\zeta}_1 + \frac{1}{2}\sigma_{1max}^2. \tag{2.26}$$

Similarly,

$$\tilde{\zeta}_2^T\underline{\delta}_2\sigma_2(t) \le \frac{1}{2}\tilde{\zeta}_2^T\bar{\delta}_2\bar{\delta}_2^T\tilde{\zeta}_2 + \frac{1}{2}\sigma_{2max}^2. \tag{2.27}$$

Substituting (2.26) and (2.27) into (2.25), we get

$$\dot{\mathcal{L}} \le -\frac{1}{2}\tilde{\zeta}_1^T\bar{\delta}_1\bar{\delta}_1^T\tilde{\zeta}_1 - \frac{1}{2}\tilde{\zeta}_2^T\bar{\delta}_2\bar{\delta}_2^T\tilde{\zeta}_2 + \frac{1}{2}(\sigma_{1max}^2 + \sigma_{2max}^2). \tag{2.28}$$

From (2.28) we can conclude that $\dot{\mathcal{L}} < 0$ when one of the following conditions holds

$$\|\tilde{\zeta}_1\| > \sqrt{\frac{\sigma_{1max}^2 + \sigma_{2max}^2}{\lambda_{min}(\bar{\delta}_1\bar{\delta}_1^T)}}, \tag{2.29}$$

or

$$\|\tilde{\zeta}_2\| > \sqrt{\frac{\sigma_{1max}^2 + \sigma_{2max}^2}{\lambda_{min}(\bar{\delta}_2\bar{\delta}_2^T)}}. \tag{2.30}$$

According to Lyapunov theory, it yields that the weight estimation errors for both critic networks are UUB.

Remark 2.6 The weight matrices are usually updated through certain renewal equations, and from (2.29) and (2.30), we can draw that the approximation weight error will asymptotically converge to zero as $\nu_\epsilon \to \infty$.

Theorem 2.7 *Consider the system (2.6). The weight updating laws for critic networks are given by (2.21). Then the obtained policies (2.18) can force system states X to be UUB.*

Proof In order to discuss the stability of closed-loop system, the derivative of $\mathcal{V} = \mathcal{V}_1^* + \mathcal{V}_2^*$ is considered as

$$\begin{aligned} \dot{\mathcal{V}} =&(\nabla \mathcal{V}_1^*)^T (f(X) + \boldsymbol{g}(X)\hat{u} + \kappa(X)\hat{v}) \\ &+ (\nabla \mathcal{V}_2^*)^T (f(X) + \boldsymbol{g}(X)\hat{u} + \kappa(X)\hat{v}). \end{aligned} \tag{2.31}$$

Recalling (2.13) and (2.14), we have

$$\begin{aligned} \nabla \mathcal{V}_1^{*T} f(X) =& - X^T \mathcal{Q}_1 X + \frac{1}{4} \nabla \mathcal{V}_1^{*T} \boldsymbol{g}(X) \mathcal{R}_{11}^{-1} \boldsymbol{g}^T(X) \nabla \mathcal{V}_1^* \\ &- \frac{1}{4} \nabla \mathcal{V}_2^{*T} \kappa(X) \mathcal{R}_{22}^{-1} \mathcal{R}_{12} \mathcal{R}_{22}^{-1} \kappa^T(X) \nabla \mathcal{V}_2^* \\ &+ \frac{1}{2} \nabla \mathcal{V}_1^{*T} \kappa(X) \mathcal{R}_{22}^{-1} \kappa^T(X) \nabla \mathcal{V}_2^*, \end{aligned} \tag{2.32}$$

and

$$\begin{aligned} \nabla \mathcal{V}_2^{*T} f(X) =& - X^T \mathcal{Q}_2 X + \frac{1}{4} \nabla \mathcal{V}_2^{*T} \kappa(X) \mathcal{R}_{22}^{-1} \kappa^T(X) \nabla \mathcal{V}_2^* \\ &- \frac{1}{4} \nabla \mathcal{V}_1^{*T} \boldsymbol{g}(X) \mathcal{R}_{11}^{-1} \mathcal{R}_{21} \mathcal{R}_{11}^{-1} \boldsymbol{g}^T(X) \nabla \mathcal{V}_1^* \\ &+ \frac{1}{2} \nabla \mathcal{V}_2^{*T} \boldsymbol{g}(X) \mathcal{R}_{11}^{-1} \boldsymbol{g}^T(X) \nabla \mathcal{V}_1^*. \end{aligned} \tag{2.33}$$

For $\epsilon = 1$, we can obtain $\dot{\mathcal{V}}_1^*$ as

$$\begin{aligned} \dot{\mathcal{V}}_1^* =& - X^T \mathcal{Q}_1 X - \frac{1}{4} \nabla \mathcal{V}_1^{*T} \boldsymbol{g}(X) \mathcal{R}_{11}^{-1} \boldsymbol{g}^T(X) \nabla \mathcal{V}_1^* \\ &- \frac{1}{4} \nabla \mathcal{V}_2^{*T} \kappa(X) \mathcal{R}_{22}^{-1} \mathcal{R}_{12} \mathcal{R}_{22}^{-1} \kappa^T(X) \nabla \mathcal{V}_2^* \\ &- \nabla \mathcal{V}_1^{*T} (\boldsymbol{g}(X)(u^* - \hat{u}) + \kappa(X)(v^* - \hat{v})). \end{aligned} \tag{2.34}$$

According to (2.15) and (2.16) we have

$$\begin{aligned} \dot{\mathcal{V}}_1^* =& - X^T \mathcal{Q}_1 X - \frac{1}{4} \nabla \mathcal{V}_1^{*T} \boldsymbol{g}(X) \mathcal{R}_{11}^{-1} \boldsymbol{g}^T(X) \nabla \mathcal{V}_1^* \\ &- \frac{1}{4} \nabla \mathcal{V}_2^{*T} \kappa(X) \mathcal{R}_{22}^{-1} \mathcal{R}_{12} \mathcal{R}_{22}^{-1} \kappa^T(X) \nabla \mathcal{V}_2^* \\ &+ \frac{1}{2} ((\nabla \xi_1(X))^T \zeta_1^* + \nabla o_1)^T \Big(\boldsymbol{g}(X) \mathcal{R}_{11}^{-1} \boldsymbol{g}^T(X) \\ &\times ((\nabla \xi_1^T(X))^T \tilde{\zeta}_1 + \nabla o_1) + \kappa(X) \mathcal{R}_{22}^{-1} \kappa^T(X) \\ &\times ((\nabla \xi_2^T(X))^T \tilde{\zeta}_2 + \nabla o_2) \Big). \end{aligned} \tag{2.35}$$

Due to Assumption 2.2 and Theorem 2.5, we obtain that

$$\begin{aligned}\dot{\mathcal{V}}_1^* \leq & - X^T \mathcal{Q}_1 X - \frac{1}{4} \nabla \mathcal{V}_1^{*T} \boldsymbol{g}(X) \mathcal{R}_{11}^{-1} \boldsymbol{g}^T(X) \nabla \mathcal{V}_1^* \\ & - \frac{1}{4} \nabla \mathcal{V}_2^{*T} \kappa(X) \mathcal{R}_{22}^{-1} \mathcal{R}_{12} \mathcal{R}_{22}^{-1} \kappa^T(X) \nabla \mathcal{V}_2^* + \theta_1,\end{aligned} \tag{2.36}$$

where the positive constant θ_1 denotes the bound of the term $\frac{1}{2}((\nabla\xi_1(X))^T \zeta_1^* + \nabla o_1)^T \Big(\boldsymbol{g}(X)\mathcal{R}_{11}^{-1}\boldsymbol{g}^T(X)((\nabla\xi_1^T(X))^T\tilde{\zeta}_1 + \nabla o_1) + \kappa(X)\mathcal{R}_{22}^{-1}\kappa^T(X)((\nabla\xi_2^T(X))^T\tilde{\zeta}_2 + \nabla o_2)\Big)$. As $\mathcal{R}_{11}$, $\mathcal{R}_{12}$ and $\mathcal{R}_{22}$ are symmetric positive definite, we have

$$\begin{aligned}& \frac{1}{4} \nabla \mathcal{V}_2^{*T} \kappa(X) \mathcal{R}_{22}^{-1} \mathcal{R}_{12} \mathcal{R}_{22}^{-1} \kappa^T(X) \nabla \mathcal{V}_2^* \\ & + \frac{1}{4} \nabla \mathcal{V}_1^{*T} \boldsymbol{g}(X) \mathcal{R}_{11}^{-1} \boldsymbol{g}^T(X) \nabla \mathcal{V}_1^* > 0.\end{aligned} \tag{2.37}$$

Furthermore, we attain

$$\dot{\mathcal{V}}_1^* \leq -X^T \mathcal{Q}_1 X + \theta_1 \leq -\lambda_{min}(\mathcal{Q}_1)\|X\|^2 + \theta_1. \tag{2.38}$$

Similarly, for $\epsilon = 2$, it yields that

$$\dot{\mathcal{V}}_2^* \leq -X^T \mathcal{Q}_2 X + \theta_2 \leq -\lambda_{min}(\mathcal{Q}_2)\|X\|^2 + \theta_2, \tag{2.39}$$

where the definition of θ_2 is similar to that of θ_1. Then it can be concluded that $\dot{\mathcal{V}} < 0$ when the following inequality is satisfied

$$\|X\| > \max\left\{\sqrt{\frac{\theta_1}{\lambda_{min}(\mathcal{Q}_1)}}, \sqrt{\frac{\theta_2}{\lambda_{min}(\mathcal{Q}_2)}}\right\} \triangleq \Theta. \tag{2.40}$$

Thus with the proposed control policies (2.18), the system state N is UUB with the bound Θ. This completes the proof.

Remark 2.8 From Theorems 2.5 and 2.7, we can conclude that under the obtained control policies the states of the system X and the critic weight error $\tilde{\zeta}_\epsilon$ are ultimately uniformly bounded.

Remark 2.9 According to the clinical requirements, the specific value of the cost function is finalised. Transformation is implemented from the mathematical mechanism model to the solvable affine model. Subsequently, the chapter solve the optimal control problem that means minimum dose of medicine can realize the best therapeutic effect.

2.4 Simulation and Numerical Experiments

To verify the proposed method in the previous section, a simulation is given as followed.

2.4.1 *States Analysis on Tumor Cell Growth*

According to clinical medical statistics borrowed from the literature [37], the specific parameters of the dynamic models are presented as Table 2.2.

According to (2.5) and Table 2.2, we construct the model (2.41)

$$
\begin{aligned}
\dot{N}_T(t) =& 0.00431 N_T(t)(1 - 1.02 \times 10^{-9})N_T(t)) \\
& - 6.41 \times 10^{-11} N_T(t) N_H(t) \\
& - 0.08 N_{CD}(t) N_T(t) - (1 - e^{-u(t)}) \\
\dot{N}_H(t) =& 0.33 + \frac{0.0125 N_H(t) N_T^2(t)}{2.02 \times 10^7 + N_T^2(t)} + \frac{0.125 N_H(t) N_{ID}(t)}{2 \times 10^7 + N_{ID}(t)} \\
& - 3.42 \times 10^{-6} N_T(t) N_H(t) - (1 - e^{-u(t)}) \\
& - 0.204 N_H(t) - 3.42 \times 10^{-6} N_{CD}(t) N_H(t) \\
\dot{N}_{CD}(t) =& u(t) - 0.1 N_{CD}(t) \\
\dot{N}_{ID}(t) =& v(t) - N_{ID}(t)
\end{aligned}
\tag{2.41}
$$

The initial state of tumor cells $N_1(t)$ and immune cells $N_2(t)$ in a patient and follow a certain chemotherapy and immunotherapy regimen. Correspondingly, $N_3(t)$ and $N_4(t)$ respectively denote the concentrations of chemotherapy and immunotherapy. And we can get the following curves on systems states tumor cells, immune cells, chemotherapy and immunotherapy drugs as shown in Fig. 2.1. Initial value is set as $X_0 = \begin{bmatrix} 20 \ 10 \ 8 \ 6 \end{bmatrix}^T$.

Table 2.2 Concentration variation on immune cells, tumor cells, chemotherapeutic drug and immunoagents

Parameter	Estimated value	Units	Parameter	Estimated value	Units
ι_1	0.00431	day^{-1}	ϱ_1	1.02×10^{-9}	cell^{-1}
σ_1	6.41×10^{-11}	cell^{-1}	δ_1	0.08	day^{-1}
$\mathfrak{D}$	0.204	day^{-1}	σ_2	3.42×10^{-6}	cell^{-1}
δ_2	2×10^{-11}	day^{-1}	υ	0.0125	day^{-1}
ς	0.125	day^{-1}	ν	2.02×10^7	cell^2
ϑ	2×10^7	cell	φ_1	0.1	day^{-1}
φ_2	1	day^{-1}			

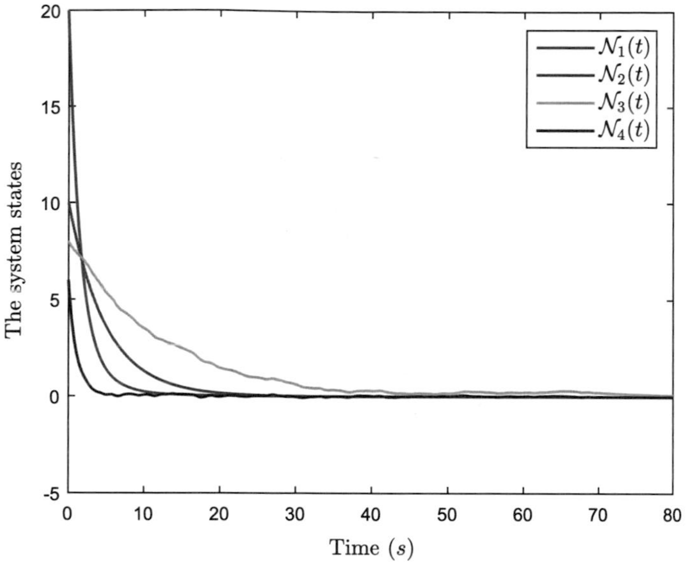

Fig. 2.1 The curves of system states

It is obviously that the control policies can stabilize the nonlinear system and make the system states tend to zero which means that the closed-system is stable and the control method is effective. Retrospect the original problem that the key is to minimize cancer cells and reduce therapy toxicity as possible.

2.4.2 Weight Analysis of Control Policies

The weights ζ_ϵ^* of the control policies $u(t)$ and $v(t)$ can be estimated through the value function $\hat{\mathcal{V}}_\epsilon^* = (\hat{\zeta}_\epsilon)^T \xi_\epsilon(X)$ in (2.16), and the performance index is shown as (2.6) with $\mathcal{Q}_1 = I_{4\times4}$, $\mathcal{Q}_2 = 5\mathcal{Q}_1$, $\mathcal{R}_{11} = \mathcal{R}_{22} = 1$, $\mathcal{R}_{12} = \mathcal{R}_{21} = 2$. The initialize weights are set as $[-0.25, -0.25, -1, -0.25]^T$. The selected activation function is selected as $[\zeta_{11\to15}^T, \zeta_{16\to18}^T, \zeta_{19\to10}^T]$, where $\zeta_{11\to15} = [N_1^2(t), N_1(t)N_2(t), N_1(t)N_3(t), N_1(t)N_4(t), N_2^2(t)]$ and $\zeta_{16\to18} = [N_2(t)N_3(t), N_2(t)N_4(t), N_3^2(t)]$ and $\zeta_{19\to10} = [N_3(t)N_4(t), N_4^2(t)]$

According to Fig. 2.2, we can conclude that the proposed optimal control demonstrated a shorter convergence time than that without taking optimal control, where the former needs only $10s$, but the later may be $38s$, which draws the superiority of the proposed method.

In Fig. 2.3, we can obtain the less doses of the drugs is another advantage compared with that without taking optimal control. Taking comprehensive consideration of Figs. 2.2 and 2.3, we can draw a conclusion that the adopted algorithm can not only

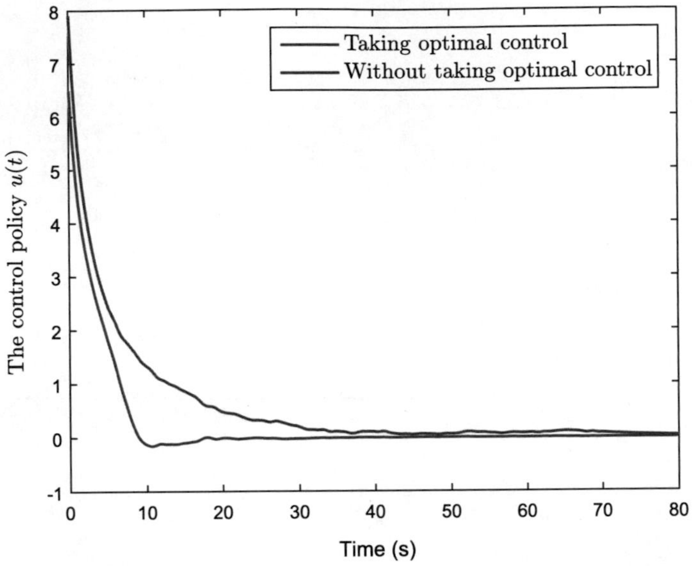

Fig. 2.2 Optimal control policies $u(t)$

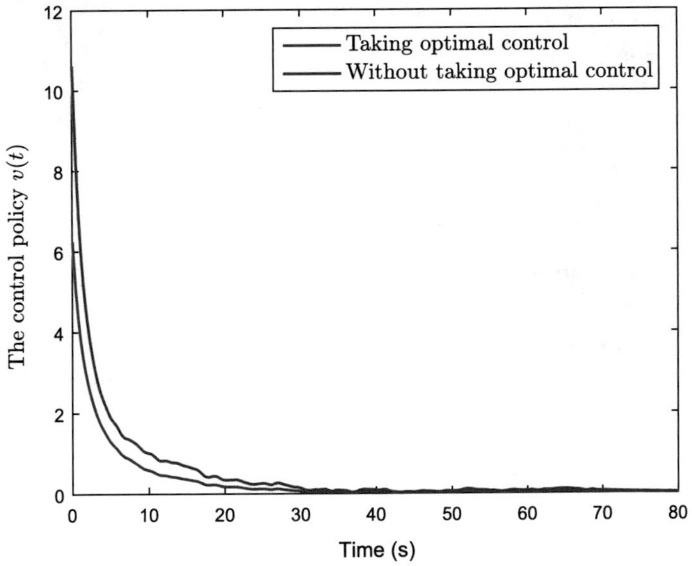

Fig. 2.3 Optimal control policies $v(t)$

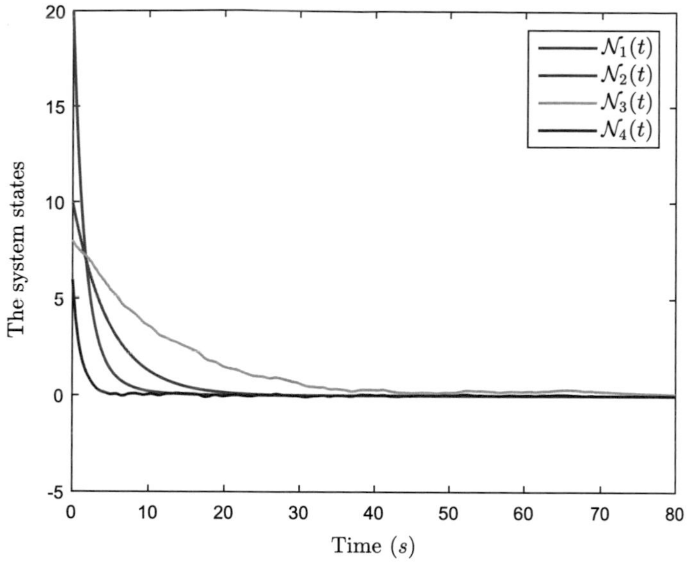

Fig. 2.4 The curves of system states

decrease the convergence time but also reduce doses of chemotherapy drugs and immune agents, and patients will benefit from for the minimal toxicity and shorter response time.

When the initialize state is set as $[-0.5, -0.1, -1, -0.4]^T$, and the other parameters are unaltered, we give another set of figures as Figs. 2.4, 2.5 and 2.6. In Figs. 2.5 and 2.6, there exist more obvious advantages for the proposed algorithms over that without taking optimal control in response time and control policies,and we can conclude that effectiveness of the control method does not vary in the different initial weights.

2.5 Conclusion

This chapter has introduced adaptive dynamic programming into solving the optimal control policies of tumor cells growth model and realized objective of minimizing tumor cells with the minimum dose of chemotherapeutic and immunotherapeutic drugs. As is known, the negative effect caused by chemotherapy and immunotherapy must be reduced for the reasonable treatment plan extracted from the optimal control behavior. Convergence properties have been proved to be guaranteed through Lyapunov theory. Meanwhile, states of the system and critic error have been demonstrated to be ultimately uniformly bounded. Simulations have been given to verify

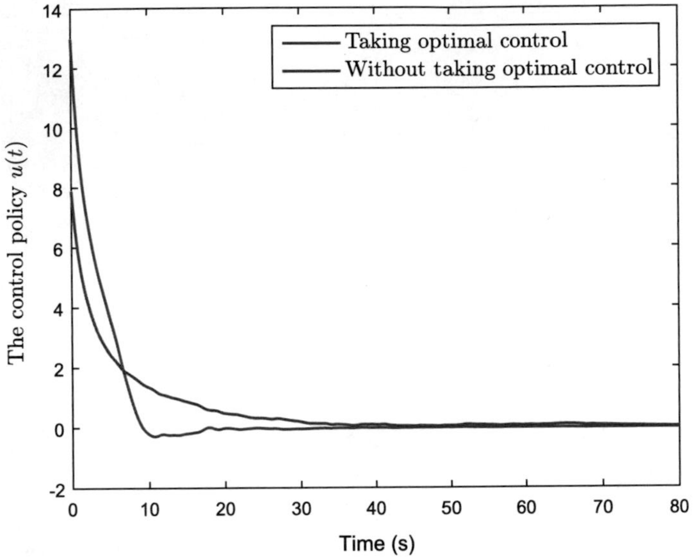

Fig. 2.5 Optimal control policies $u(t)$

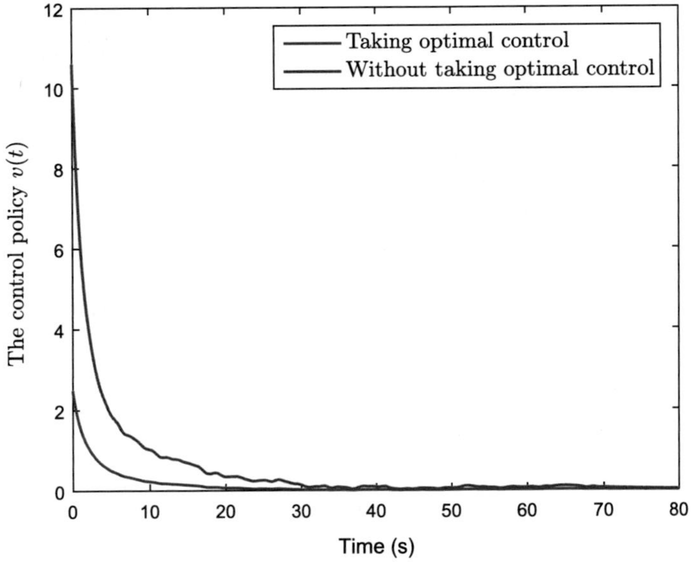

Fig. 2.6 Optimal control policies $v(t)$

rationality of the proposed method. In the future work, we will further investigate the medical frontier topics and propose adaptive therapeutic methods to solve these issues by employing ADP approach.

References

1. de Pillis LG, Gu W, Radunskaya AE (2006) Mixed immunotherapy and chemotherapy of tumors: modeling, applications and biological interpretations. J Theor Biol 238(4):841–862
2. Ogunmadeji B, Yusuf TT (2018) Optimal control strategy for improved cancer biochemotherapy outcome. Int J Sci Eng Res 9(12):583–600
3. Liang H, Liu G, Zhang H, Huang T (2021) Neural-network-based event-triggered adaptive control of nonaffine nonlinear multiagent systems with dynamic uncertainties. IEEE Trans Neural Netw Learn Syst 32(5):2239–2250
4. Sun J, Zhang H, Wang Y, Sun S (2022) Fault-tolerant control for stochastic switched IT2 fuzzy uncertain time-delayed nonlinear systems. IEEE Trans Cybernet 52(2):1335–1346
5. Lodhi I, Ahmad I, Uneeb M, Liaquat M (2019) Nonlinear Control for Growth of Cancerous Tumor Cells. IEEE Access 7:177628–177636
6. Wang J, Huang M, Chen S, Luo Y, Shen S, Du X (2021) Nanomedicine-mediated ubiquitination inhibition boosts antitumor immune response via activation of dendritic cells. Nano Res 14:3900–3906
7. Kuznetsov VA, Makalkin IA (1992) Bifurcation analysis of a mathematical model of the interaction of cytotoxic lymphocytes with tumor cells. The effect of immunologic amplification of tumor growth and its interconnection with other anomolous phenomena of oncoimmunology. Biofizika 37(6):1063–1070
8. Zhang H, Cui L, Luo Y (2013) Near-optimal control for nonzero-sum differential games of continuous-time nonlinear systems using single-network ADP. IEEE Trans Cybernet 43(1):206–216
9. Fan QY, Xu S, Xu B, Qiu J (2022) Simplified prescribed performance tracking control of uncertain nonlinear systems. Sci China Inf Sci 65:189204
10. Starr WA, Ho YC (1969) Nonzero-sum differential games. J Optim Theory Appl 3(2):184–206
11. Su H, Zhang H, Jiang H, Wen Y (2020) Decentralized event-triggered adaptive control of discrete-time nonzero-sum games over wireless sensor-actuator networks with input constraints. IEEE Trans Neural Netw Learn Syst 31(10):4254–4266
12. Wei Q, Zhu L, Song R, Zhang P, Liu D, Xiao J (2022) Model-free adaptive optimal control for unknown nonlinear multiplayer nonzero-sum game. IEEE Trans Neural Netw Learn Syst 33(2):879–892
13. Song R, Lewis FL, Wei Q (2017) Off-policy integral reinforcement learning method to solve nonlinear continuous-time multiplayer nonzero-sum games. IEEE Trans Neural Netw Learn Syst 28(3):704–713
14. Zhang K, Su R, Zhang H, Tian Y (2021) Adaptive resilient event-triggered control design of autonomous vehicles with an iterative single critic learning framework. IEEE Trans Neural Netw Learn Syst 32(12):5502–5511
15. Yang D, Li T, Xie X, Zhang H (2020) Event-triggered integral sliding-mode control for nonlinear constrained-input systems with disturbances via adaptive dynamic programming. IEEE Trans Syst Man Cybernet: Syst 50(11):4086–4096
16. Wei Q, Li H, Yang X, He H (2021) Continuous-time distributed policy iteration for multicontroller nonlinear systems. IEEE Trans Cybernet 51(5):2372–2383
17. Zhao B, Liu D, Luo C (2020) Reinforcement learning-based optimal stabilization for unknown nonlinear systems subject to inputs with uncertain constraints. IEEE Trans Neural Netw Learn Syst 31(10):4330–4340

18. Zhang T, Su G, Qing C, Xu X, Cai B, Xing X (2021) Hierarchical lifelong learning by sharing representations and integrating hypothesis. IEEE Trans Syst Man Cybernet: Syst 51(2):1004–1014
19. Narayanan V, Sahoo A, Jagannathan S, George K (2019) Approximate optimal distributed control of nonlinear interconnected systems using event-triggered nonzero-sum games. IEEE Trans Neural Netw Learn Syst 30(5):1512–1522
20. Zhong X, He H (2017) An event-triggered ADP control approach for continuous-time system with unknown internal states. IEEE Trans Cybernet 47(3):683–694
21. Sun J, Zhang H, Wang Y, Shi Z (2022) Dissipativity-based fault-tolerant control for stochastic switched systems with time-varying delay and uncertainties. IEEE Trans Cybernet 52(10):10683–10694
22. Zhang K, Zhang H, Mu Y, Liu C (2021) Decentralized tracking optimization control for partially unknown fuzzy interconnected systems via reinforcement learning method. IEEE Trans Fuzzy Syst 29(4):917–926
23. Li T, Yang D, Xie X, Zhang H (2022) Event-triggered control of nonlinear discrete-time system with unknown dynamics based on HDP(λ). IEEE Trans Cybernet 52(7):6046–6058
24. Mu C, Wang K, Ni Z (2022) Adaptive learning and sampled-control for nonlinear game systems using dynamic event-triggering strategy. IEEE Trans Neural Netw Learn Syst 33(9):4437–4450
25. Mu C, Wang K, Qiu T (2022) Dynamic event-triggering neural learning control for partially unknown nonlinear systems. IEEE Trans Cybernet 52(4):2200–2213
26. Zhang H, Qin C, Jiang B, Luo Y (2014) Online adaptive policy learning algorithm for H_∞ state feedback control of unknown affine nonlinear discrete-time systems. IEEE Trans Cybernet 44(12):2706–2718
27. Zhang Q, Zhao D (2019) Data-based reinforcement learning for nonzero-sum games with unknown drift dynamics. IEEE Trans Cybernet 49(8):2874–2885
28. Chen G, Yao D, Zhou Q, Li H, Lu R (2022) Distributed event-triggered formation control of usvs with prescribed performance. J Syst Sci Complex 35(3):820–838
29. Zhang H, Su H, Zhang K, Luo Y (2019) Event-triggered adaptive dynamic programming for non-zero-sum games of unknown nonlinear systems via generalized fuzzy hyperbolic models. IEEE Trans Fuzzy Syst 27(11):2202–2214
30. Wei Q, Song R, Liao Z, Li B, Lewis FL (2020) Discrete-time impulsive adaptive dynamic programming. IEEE Trans Cybernet 50(10):4293–4306
31. Lewis FL, Vrabie DL, Syrmos VL (2012) Optimal control. Wiley, Hoboken, New Jersey, USA
32. Başar T, Olsder GJ (1999) Dynamic noncooperative game theory. SIAM, Philadelphia, PA, USA
33. Vamvoudakis KG (2014) Event-triggered optimal adaptive control algorithm for continuous-time nonlinear systems. IEEE/CAA J Autom Sinica 1(3):282–293
34. Vamvoudakis KG, Lewis FL (2010) Online actor-critic algorithm to solve the continuous-time infinite horizon optimal control problem. Automatica 46(5):878–888
35. Wang D, He H, Zhong X, Liu D (2017) Event-driven nonlinear discounted optimal regulation involving a power system application. IEEE Trans Ind Electron 64(10):8177–8186
36. Zhang H, Zhang J, Yang G-H, Luo Y (2015) Leader-based optimal coordination control for the consensus problem of multiagent differential games via fuzzy adaptive dynamic programming. IEEE Trans Fuzzy Syst 23(1):152–163
37. Sharma S, Samanta GP (2016) Analysis of the dynamics of a tumor-immune system with chemotherapy and immunotherapy and quadratic optimal control. Differ Equ Dyn Syst 24(2):149–171

Chapter 3
Optimal Regulation Strategy for Nonzero-Sum Games of the Immune System Using Adaptive Dynamic Programming

3.1 Introduction

As the rapid increase of tumor patients, immunotherapies integrated with multi-pronged approaches are being burgeoning for treatment of cancers with specific forms, especially for poorly immunogenic tumors as [1]. The original intention of immunotherapy is fighting cancer cells with their own lethality of immune cells. AIDS as a typical immunodeficiency syndrome caused by failure of immune response tends to be attributed to weakened immune levels, however, Natural killer cell population determine whether shutdown of immune system, once the activate immune system can not be suspended from and produce cytokines [2], which is regarded as an overreaction of the immune system such as COVID-19. Thus, the Nash equilibrium between the tumor cells and the immune cell population needs to be solved through optimal regulation based on specific learning method, and optimal control scheme is firstly brought into this field with its unique superiority, what's more, nonzero-sum games-based ADP enjoys meliority and practicability.

Decision and estimation on unknown nonlinearity existed so extensively in fields of engineering practice, medical treatments and even the social sciences, such that literature [3] firstly proposed the evaluation of the designed S-Box with highly nonlinearity on the basis of Chinese I-Ching philosophy. It is of great importance to make a suitable treatment decision in the field of health care where remains highly nonlinearity. To obtain an optimal mixed treatment strategy, the growth model of cell population levels was developed based on combination of immune and chemotherapy as literatures [4, 5]. When it comes to reaction of the immune system to tumor growth, a rather complicated nonlinear model of the immune system is requisite to simulate the overall aggressive combination treatment plan of immunotherapy and chemotherapy well. Thus, the process of solving the nonlinear function is hardly to be achieved unless the application of exceptionally optimized iterative algorithm such as backstepping techniques in [6], self-learning optimal regulation in [7], hierarchical lifelong learning as [8], broad learning adaptive neural control in [9] and adaptive dynamic programming, which benefits from its adaptive capability and strong

J. Sun et al., *Adaptive Dynamic Programming*,
https://doi.org/10.1007/978-981-99-5929-7_3

autonomous iterative learning ability [10, 11]. Whether backstepping or adaptive dynamic programming both could guarantee the control objective would be achieved, and unknown nonlinear function matched the value of successive searching approximation through neural networks or fuzzy control as literatures [12–15].

$\mathcal{H}_\infty$ control enjoys excellent disturbance suppression while minimizing performance index and it is recognized as a typical two-player zero-sum problem, which can be equivalent to solving algebraic Riccati equations, and it is generally applied into linear dynamics systems, of course, systems with quadratic performance index could be actually solved such as literature [16]. Meanwhile, the familiar Hamilton-Jacobi-Isaacs is perceived as an effective medium in dealing with systems considering inherent nonlinearity, such as unknown mechanical parameters in [17], which is difficult to achieve using conventional methods for absence of exact system parameters. The mainstream analysis of ADP is seeking optimal control strategy integrated with solution to Bellman functions without information of system dynamics, which has ascended to the core methodologies of optimization and artificial intelligence. When it comes to actual models, control constraint has been definitely considered as [18–20], thus the chapter mainly focuses on dynamic model of the immune system which limits the single injection of drugs to an intervention level, and the optimal control scheme is transformed into constrained control which needs to take a discounted factor into account, avoiding infinite time dimension effectively, which will lead to development of optimal constrained control policy.

Model-free adaptive control was developed to obtain optimal control strategy without knowledge of exact system parameters as literatures [21–23], and multiple neural networks were constructed to achieve multi-objective approximation or optimization control process. Research with respect to multiple networks has been extended to multitudinous actor-critic constructions. A tremendous amount of practical application scenarios need multiple controllers, each of which minimizes its individual performance function as nonzero-sum problem. As elaborated in nonzero-sum game theory, the control objective was minimizing the individual performance function and maintaining stability to yield a Nash equilibrium in [24]. As literature [25], saddle point of the Nash equilibrium was explored throughout the nonzero-sum games-based optimization iterative process using ADP, even if there was no feasible saddle point, optimum was realized through mixed optimal control scheme iteratively, and the latter is of universal significance for conditions that are uneasy to satisfy in practical applications, The local optimal problem exits extensively which was firstly effectively avoided through fault-tolerant adaptive multigradient recursive reinforcement learning as [9]. To seek the solution to Nash equilibrium, the simultaneous algebraic Riccati or Hamilton-Jacobi-Isaacs functions require solving for nonlinear systems, which leads to "curse of dimensionality" with huge amount of computation, especially for multitudinous actor-critic constructions suffering from higher computational burden by many multiples, such as a double-loop policy iteration in [26]. According to the reason described above, the chapter adopts compromise acceptable actor-critic neural networks with appropriate dimensions, effectively realizing the transformation process from value iterative to cost function.

Value and policy iterations generally constitute the whole iterative methods, and begin with an semidefinite function or admissible control law accordingly. With applications of ADP to solve the optimal control strategy for both continuous [27, 28] or discrete-time systems [29, 30], however, traditional ADP can not satisfy the physical application in the immune system considering the mixed treatment strategy with chemotherapy drugs and immunotherapy, improving matters somewhat by nonzero-sum games-based ADP. There are seldom any literatures on nonzero-sum games-based ADP method for solving optimal regulation schemes of the immune system, let alone considering optimal constrained control, policy iterations, tumor regression and mixed control strategy of chemotherapy and immunotherapy, scilicet the cost function approaching covers minimization of the tumor cells, chemotherapy drugs and immunotherapy drugs, simultaneously.

3.2 Establishment of Mathematical Model

This part mainly introduces the mathematical growth model of tumor cells, which considers the influence of external factors such as chemotherapy drugs and immunotherapy on the tumor cells, mutual effect between two types of cells. In the following model, $Tu(t)$ represents the amount of tumor cells, $Im(t)$ denotes the number of immune cells, and $Che(t)$, $Im_{py}(t)$ depicts the concentrations of chemotherapy drugs and immunotherapy drugs in the bloodstream, respectively.

3.2.1 *Growth Model of Tumor Cells*

Individually considering the natural growth law of tumor cells without the relationship with immune cells and any external effect on them, the growth law of tumor cells is subject to logical growth.

$$Tu(t+1) = Tu(t) + C_1 Tu(t)(1 - C_2 Tu(t)). \tag{3.1}$$

But when it comes to the interaction between immune cells and tumor cells, the direct killing of cells by chemotherapeutic drugs, and the growth model of tumor cells can be revised to:

$$\begin{aligned} Tu(t+1) = & Tu(t) + C_1 Tu(t)(1 - C_2 Tu(t)) \\ & - C_{Im,Tu} Tu(t) Im(t) - C_{Che,Tu} Tu(t) Che(t), \end{aligned} \tag{3.2}$$

where the specifications of parameters are demonstrated as Table 3.1.

Table 3.1 Parameter specifications of the tumor cells

Parameters	Relationships with the tumor cells
C_1	Intrinsic growth rate irrespective of the immune cells and chemotherapy drugs
C_2	Maximum capacity for interaction between tumor cells ignoring effects of immune cells and chemotherapy drugs
$C_{Im,Tu}$	Growth rate when the tumor cells are inactivated to attack from the immune cells
$C_{Che,Tu}$	Stress response coefficient of the tumor cells to chemotherapy drugs

3.2.2 Growth Model of Immune Cells

Considering the natural growth law of immune cells simply, we assume that a fixed number of immune cells are produced in a unit of time and that these cells have an inevitable life cycle.

$$Im(t+1) = Im(t) + C_3 - C_{Im,d}Im(t) \tag{3.3}$$

The tumor cells in the body can stimulate the growth of immune cells, which shows a positive non-linear change by (3.4).

$$\Delta_{im} = \frac{\alpha_1 Tu(t)^2 Im(t)}{\beta_1 + Tu(t)^2}. \tag{3.4}$$

In immunotherapy, the addition of immune agents can produce an immune response, which leads to the non-linear growth of immune cells.

$$\Delta_{Im_{py}} = \frac{\alpha_2 Tu(t) Im_{py}(t)}{\beta_2 + Im_{py}(t)}. \tag{3.5}$$

Simultaneously, in the struggle between immune cells and tumor cells, immune cells themselves can also cause losses,

$$\Delta_{C_{Tu,Im}} = -C_{Tu,Im}Tu(t)Im(t). \tag{3.6}$$

and in chemotherapy, chemotherapeutic drugs can also cause damage to immune cells.

$$\Delta_{C_{Che,Im}} = -C_{Che,Im}Che(t)Im(t). \tag{3.7}$$

Combined (3.3)–(3.7), and then (3.8) can be obtained.

Table 3.2 Parameter specifications of the immune cells

Parameters	Relationships with the immune cells
C_3	Constant inflow rate
$C_{Im,d}$	Natural decay rate without any external action
α_1	Maximum rate of recruitment caused by the tumor cells
β_1	Steepness coefficient caused by the tumor cells
α_2	Maximum rate of tumor cells caused by immunotherapy drug
β_2	Steepness coefficient caused by immunotherapy drug
$C_{Che,Im}$	Stress response coefficient to chemotherapy drug
$C_{Tu,Im}$	Reaction rate of tumor cells to the immune cells

$$\begin{aligned} Im(t+1) &= Im(t) + C_3 - C_{Im,d}Im(t) + \Delta_{im} + \Delta_{Im_{py}} \\ &\quad + \Delta_{C_{Tu,Im}} + \Delta_{C_{Che,Im}} \\ &= Im(t) + C_3 - C_{Im,d}Im(t) + \frac{\alpha_1 Tu(t)^2 Im(t)}{\beta_1 + Tu(t)^2} \\ &\quad + \frac{\alpha_2 Tu(t) Im_{py}(t)}{\beta_2 + Im_{py}(t)} - C_{Tu,Im}Tu(t)Im(t) \\ &\quad - C_{Che,Im}Che(t)Im(t). \end{aligned} \tag{3.8}$$

Parameter elucidation of immune cells are outlined as Table 3.2.

3.2.3 Drug Attenuation Model

We assume that at some point after the injection of a chemotherapy drug, the concentration of the drug in the body will decrease exponentially. To guarantee the effectiveness of the treatment, we add chemotherapy drugs to the body, simultaneously.

$$Che(t+1) = Dr_{Che}(t) - e^{-\gamma_1}Che(t). \tag{3.9}$$

Similarly, we can obtain the attenuation model of the immunoagents:

$$Im_{py}(t+1) = Dr_{Im}(t) - e^{-\gamma_2}Im_{py}(t). \tag{3.10}$$

where injected at t, $Dr_{Che}(t)$ and $Dr_{Im}(t)$ denotes concentrations of the chemotherapy drugs and immunoagents separately. γ_1 and γ_2 is the decay rates of the chemotherapy drugs and immunoagents.

3.2.4 The Design of the Optimization Problem

Combined with the contents of (A), (B) and (C), we finally obtain the mathematical model affecting the growth of tumor cells:

$$\begin{cases} Tu(t+1) = Tu(t) + C_1 Tu(t)(1 - C_2 Tu(t)) \\ -C_{Im,Tu} Tu(t) Im(t) - C_{Che,Tu} Tu(t) Che(t) \\ Im(t+1) = Im(t) + C_3 - C_{Im,d} Im(t) \\ +\frac{\alpha_1 Tu(t)^2 Im(t)}{\beta_1 + Tu(t)^2} + \frac{\alpha_2 Tu(t) Im_{py}(t)}{\beta_2 + Im_{py}(t)} \\ -C_{Tu,Im} Tu(t) Im(t) - C_{Che,Im} Che(t) Im(t) \\ Che(t+1) = Dr_{Che}(t) - e^{-\gamma_1} Che(t) \\ Im_{py}(t+1) = Dr_{Im}(t) - e^{-\gamma_2} Im_{py}(t). \end{cases} \tag{3.11}$$

Given that $Tu(t)$, $Im(t)$ are biomass, and $Che(t)$, $Im_{py}(t)$ are the drug concentrations in the bloodstream,

$$Tu(t), Im(t), Che(t), Im_{py}(t) \geq 0, \forall t > 0. \tag{3.12}$$

And all parameters in the model are non-negative:

$$\begin{aligned} &C_1; C_2; C_3; C_{Im,Tu}; C_{Che,Tu}; C_{Im,d}; C_{Tu,Im}; C_{Che,Im} \\ &\alpha_1; \alpha_2; \beta_1; \beta_2; \gamma_1; \gamma_2 \geq 0, \forall t > 0. \end{aligned} \tag{3.13}$$

When we qualitatively analyze the problem that how to minimize the residual tumor cell population in the bloodstream on the premise of using as few drugs as possible, including chemotherapy drugs and immunoagents. This process can be described as a quantitative mathematical expression as (3.14).

$$\begin{aligned} \min\{aTu(t)^2 + b_1 \int_0^{Dr_{Che}(t)} tanh^{-1}(\bar{U}_1^{-1} s)\bar{U}_1 R_1 ds \\ + b_2 \int_0^{Dr_{Im}(t)} tanh^{-1}(\bar{U}_2^{-1} s)\bar{U}_2 R_2 ds\}. \end{aligned} \tag{3.14}$$

It is emphasized here that the single dose of the two drugs should be limited to avoid drug poisoning. So we use a definition form with input constraints. During the whole treatment process, we can get:

$$\sum_{t=t_0}^{t_f} \lambda^t \{aTu(t)^2 + b_1 \int_0^{Dr_{Che}(t)} tanh^{-1}(\bar{U}_1^{-1}s)\bar{U}_1 R_1 ds$$
$$+ b_2 \int_0^{Dr_{Im}(t)} tanh^{-1}(\bar{U}_2^{-1}s)\bar{U}_2 R_2 ds\}, \quad (3.15)$$

where $0 < \lambda < 1$, $\bar{U}_1$ and $\bar{U}_2$ represent the maximum permissible dose of chemotherapy drug and dose of immune agents in a single injection, respectively.

3.3 The Proposed Nonzero-Sum Games-Based ADP Scheme

To solve the given problems above, we propose an aggressive treatment plan or control scheme based on nonzero-sum games-based ADP algorithm.

3.3.1 Theoretical Introduction

For a differential control system $x(t+1) = F(x(t), u(t), t))$, $x(t)$ is the state variable, $u(t)$ is the control variable, F is the transition mapping between states, and then the cost of state transition is obtained. $U(x(t), u(t), t)$, and the total cost of the whole period is $\sum_{t=t_0}^{t_f} U(x(t), u(t), t)$.

When solving a finite time problem, we can equivalent it to

$$\sum_{t=t_0}^{\infty} \lambda^t U(x(t), u(t), t), 0 < \lambda < 1. \quad (3.16)$$

In the application of Bellman's optimality principle to solve (3.1), we first stipulate $J(x(t_0)) = \sum_{t=t_0}^{\infty} \lambda^t U(x(t), u(t), t)$, and then we can obtain that

$$J^*(x(t)) = \min_{u(t)} \left\{U[x(t), u(t)] + \lambda J^*[x(t+1)]\right\}, t \in (t_0, \infty). \quad (3.17)$$

The corresponding optimal control can be solved and the form as follows.

$$u^*[x(t)] = \arg\min_{u(t)} \left\{U[x(t), u(t)] + \lambda J^*[x(t+1)]\right\}, t \in (t_0, \infty). \quad (3.18)$$

This typical solution approach is a considerable challenge for computing and storage space.

Remark 3.1 Adaptive dynamic programming as an optimize learning method is usually used to track the cost function, which is not only designed to minimize the tumor cells, but also minimum dose chemotherapy drugs and immunoagents in this chapter.

3.3.2 *Iterative ADP Algorithm*

To solve (1), we use an iterative adaptive dynamic programming algorithm, and the revised facilitate solving differential equations model.

(1) Brief interpretation of ADP algorithm
Firstly, we take a value function $K(x)$ to approximate the cost function $J(x)$. In this case, the purpose of iteration is to ensure that the approximate function approaches to the optimal value equation and obtain the optimal decision law. Namely,

$$\begin{cases} K(x) \to J^*(x) \\ \kappa \to u^*. \end{cases} \tag{3.19}$$

Secondly, in the specific solution process:
Give $K^0(\cdot) = 0$, we make

$$\kappa^0(x(t)) = \arg\min_{u(t)} \left\{U[x(t), u(t)] + \lambda K^0(x(t+1))\right\}, \tag{3.20}$$

and update the value function as

$$K^1(x(t)) = U[x(t), \kappa^0(x(t))] + \lambda K^0(x(t+1)), \tag{3.21}$$

for $i = 1, 2, 3, \ldots$, we can get

$$\kappa^i(x(t)) = \arg\min_{u(t)} \left\{U[x(t), u(t)] + \lambda K^i(x(t+1))\right\}. \tag{3.22}$$

and

$$K^{i+1}(x(t)) = \min_{u(t)} \left\{U[x(t), u(t)] + \lambda K^i(x(t+1))\right\}. \tag{3.23}$$

Thus,

$$K^{i+1}(x(t)) = U[x(t), \kappa^i(x(t))] + \lambda K^i(x(t+1)). \tag{3.24}$$

and the optimal solution is obtained when the error requirement has been adequately satisfied as condition that $K^i(x(t)) \to K^*(x(t))$ and $\left\| K^{i+1}(x(t)) - K^i(x(t+1)) \right\| \leq \varepsilon$, where i represents the number of iterations.

Algorithm : EvolutionaryADPalgorithm

Initialization :

1. A certain initial state is given randomly in the feasible region $x(t)$;
2. Set $\Lambda^0(\cdot) = 0$;
3. Specific parameters are given according to the requirements: error ϵ, discount factor λ;

IterationandUpdate :

4. $i = 0$, substitute $x(t)$ into "(3.26) = 0 ", yield $\kappa^i(t)$;
5. Plug $x(t)$ and $\kappa^i(t)$ into (3.25),and to get $x(t+1)$;
6. According to the (3.29), calculate $\Lambda^{i+1}(x(t)) = \frac{\partial U(x(t+1), \kappa^i(t))}{\partial x(t)} + \Lambda^i(x(t+1))$;
7. According to the data set $[x(t), \Lambda^{i+1}(x(t))]$, the neural network of the relationship between $x \sim \Lambda$;
8. Using the neural network obtained by "7.", the value in the same state is calculated. When $\left\| \Lambda^{i+1}(x(t)) - \Lambda^i(x(t)) \right\| \leq \epsilon$,ends; If it is not true, returns "4.";

To faster convergence to the optimal solution, we update in each iteration and value function, the control law according to the current direction of steepest descent, that is,

$$\frac{\partial K^{i+1}(x(t))}{\partial x(t)} = \frac{\partial U(x(t), \kappa^{i+1}(t))}{\partial x(t)} + \lambda [\frac{\partial x(t+1)}{\partial x(t)}]^T \frac{\partial K^i(x(t+1))}{\partial x(t+1)}, \tag{3.25}$$

$$\frac{\partial K^{i+1}(x(t))}{\partial \kappa^{i+1}(t)} = \frac{\partial U(x(t), \kappa^{i+1}(t))}{\partial \kappa^{i+1}(t)} + \lambda [\frac{\partial x(t+1)}{\partial \kappa^{i+1}(t)}]^T \frac{\partial K^i(x(t+1))}{\partial x(t+1)}. \tag{3.26}$$

Setting $\Lambda^i(x(t+1)) = \frac{\partial K^i(x(t+1))}{\partial x(t+1)}$:

(2) Modification of Model (3.11)

Compared with the traditional control strategy, we directly solve the problem proposed in this chapter by using ADP, although it is difficult to solve the model. Here, we propose a fitting idea to modify the model. Analysis on (3.11) shows that the injection of chemotherapy drugs into the body has a direct effect on tumor cells. On the other hand, immunoagents act on immune cells, which affects tumor cell populations. Throughout the whole action process, we can only consider the input of chemotherapy drugs and immunoagents at every moment as the two control inputs of the system and the state variables of the system are selected as the intermediate transition variables such as tumor cells and immune cell population.

1. The standard expressions of control variables, state variables, cost functions and so on are given as follows,

$$x(t) = Tu(t), u_1(t) = Dr_{che}(t), u_2(t) = Dr_{Im}(t). \tag{3.27}$$

$$K(x) = \sum_{t=t_0}^{\infty} \lambda^t \{ax(t)^2 + b_1 \int_0^{u_1(t)} tanh^{-1}(\bar{U}_1^{-1} s)\bar{U}_1 R_1 \, ds$$
$$+ b_2 \int_0^{u_2(t)} tanh^{-1}(\bar{U}_2^{-1} s)\bar{U}_2 R_2 \, ds\}. \tag{3.28}$$

2. The modified system model adopts the form of nonlinear affine system, namely:

$$x(t+1) = f(x(t)) - [g_1(x(t)), g_2(x(t))][u_1(t), u_2(t)]^T. \tag{3.29}$$

3. Update the optimal control law and value function:
   Let $\frac{\partial K^{i+1}(x(t))}{\partial u_1^i(t)} = 0$ and $\frac{\partial K^{i+1}(x(t))}{\partial u_2^i(t)} = 0$,

$$u_1^{i,*}(t) = \bar{U}_1 tanh(\frac{\lambda}{b_1 \bar{U}_1 R_1} g_1(x(t)) \Lambda^i(x(t+1))), \tag{3.30}$$

$$u_2^{i,*}(t) = \bar{U}_2 tanh(\frac{\lambda}{b_2 \bar{U}_2 R_2} g_2(x(t)) \Lambda^i(x(t+1))). \tag{3.31}$$

From this, we can also get

$$\frac{\partial K^{i+1}(x(t))}{\partial x(t)} = \Lambda^{i+1}(x(t)) = \lambda[\frac{\mathrm{d}f(x)}{\mathrm{d}x} - u_1^i \frac{\mathrm{d}g_1(x)}{\mathrm{d}x} - u_2^i \frac{\mathrm{d}g_2(x)}{\mathrm{d}x}]$$
$$\cdot \Lambda^i(x(t+1)) + 2ax. \tag{3.32}$$

Remark 3.2 To approximate optimal value based on optimal decision law, value iteration method is devoted to tending to the cost function $J(x)$ through value function $K(x)$.

Remark 3.3 The fitted curve is constructed according to date obtained from the original model which is uneasy to solve, and the modification of model is research objectives for replacement, considering control inputs as chemotherapy drugs and immunoagents, simultaneously.

3.3.3 Convergence Analysis

This section provides proof of the convergence of this algorithm to prove the effectiveness of the algorithm in theory. This proof is mainly derived from formulas (3.1), (3.2), and (3.3), including two lemmas and three theorems.

Lemma 3.4 *Take a control sequence* $\{\vec{Ar}^{i}(\vec{x}(t))\}$. *When it is brought into formula (1), the corresponding value function* $J^{i}_{Ar}(\vec{x})$ *is obtained. Compared with the control sequence* $\{\vec{\kappa}^{i}(\vec{x}(t))\}$ *corresponding to the minimum cost* $K^{i}(\vec{x}(t))$. *If* $J^{0}_{Ar}(\cdot) = K^{0}(\cdot) = 0$, $J^{i+1}_{Ar}(\vec{x}(t)) = U[\vec{x}(t), \vec{Ar}^{i}(\vec{x}(t))] + \lambda J^{i}_{Ar}(\vec{x}(t+1))$, *satisfying*

$$\begin{aligned} K^{i+1}(\vec{x}(t)) &= U[\vec{x}(t), \vec{\kappa}^{i}(\vec{x}(t))] + \lambda J^{i}_{Ar}(\vec{x}(t+1)) \\ &= \min_{Ar(t)} \left\{ U[\vec{x}(t), Ar(t)] + \lambda K^{i}(\vec{x}(t+1)) \right\} \end{aligned} \tag{3.33}$$

Then, $J^{i}_{Ar}(\vec{x}(t)) \geq K^{i}(\vec{x}(t))$ *for* $\forall i$.

Proof $K^{i}(\vec{x})$ is obtained by taking the minimum value equation $J^{i}_{Ar}(\vec{x})$. $\{\vec{\kappa}^{i}(\vec{x}(t))\}$ is the corresponding optimal control sequence. For the arbitrarily control sequence $\{\vec{Ar}^{i}(\vec{x}(t))\}$, the value equation $J^{i}_{Ar}(\vec{x})$ which is corresponding with the arbitrarily control sequence must not be less than $K^{i}(\vec{x})$.

Lemma 3.5 *Select a stable admissible control sequence* $\{\vec{Sa}^{i}(\vec{x}(t))\}$ *with certain restrictions and the corresponding value equation* $J^{i}_{Sa}(\vec{x})$. *For controllable system, if* $J^{0}_{Sa}(\cdot) = K^{0}(\cdot) = 0$ *and* $J^{i+1}_{Sa}(\vec{x}(t)) = U[\vec{x}(t), \vec{Sa}^{i}(\vec{x}(t))] + \lambda^{i}(\vec{x}(t+1))$, *Then* $J^{i}_{Sa}(\vec{x})$ *is bounded.*

Proof

$$\begin{aligned} J^{i+1}_{Sa}(\vec{x}(t)) &= U[\vec{x}(t), \vec{Sa}^{i}(\vec{x}(t))] + \lambda J^{i}_{Sa}(\vec{x}(t+1)) \\ &= U[\vec{x}(t), \vec{Sa}^{i}(\vec{x}(t))] + \lambda U[\vec{x}(t), \vec{Sa}^{i-1}(\vec{x}(t+1))] \\ &\quad + \lambda J^{i-1}_{Sa}(\vec{x}(t+2)) \\ &= U[\vec{x}(t), \vec{Sa}^{i}(\vec{x}(t))] + \lambda U[\vec{x}(t), \vec{Sa}^{i-1}(\vec{x}(t+1))] \\ &\quad + \lambda^{2} U[\vec{x}(t+2), \vec{Sa}^{i-2}(\vec{x}(t+2))] + \ldots \\ &\quad + \lambda^{i+1} J^{0}_{Sa}(\vec{x}(t+i+1)). \end{aligned} \tag{3.34}$$

Thus, $J^{i+1}_{Sa}(\vec{x}(t)) = \sum_{j=0}^{i} \lambda^{i} U[\vec{x}(t+j), \vec{Sa}^{i-j}(\vec{x}(t+j))]$ and $J^{i+1}_{Sa}(\vec{x}(t)) \leq \lim_{i\to\infty} \sum_{j=0}^{i} \lambda^{i} U[\vec{x}(t+j), \vec{Sa}^{i-j}(\vec{x}(t+j))]$, where $\{Sa^{i}(\vec{x})\}$ is the stable allowable control sequence, and we can get an conclusion that $0 \leq J^{i+1}_{Sa}(\vec{x}(t)) \leq \lim_{i\to\infty} \sum_{j=0}^{i} \lambda^{i} U[\vec{x}(t+j), \vec{Sa}^{i-j}(\vec{x}(t+j))] \leq C$ for given constant C. That is, $J^{i}_{Sa}(\vec{x})$ is bounded.

Theorem 3.6 *From formula (1),* $\{\vec{\kappa}^{i}(\vec{x}(t))\}$ *is the control sequence corresponding to the minimum value function* $K^{i}(\vec{x})$. *Assuming the initial state* $K^{i}(\cdot) = 0$, *it can be proved that the sequence* $\{\vec{\kappa}^{i}(\vec{x}(t))\}$ *is a monotonic non-decreasing sequence, and* $K^{i}(\vec{x}(t)) \leq K^{i+1}(\vec{x}(t))$.

Proof Define a value equation $T^i(\vec{x}(t)) : T^i(\cdot) = 0, T^{i+1}(\vec{x}(t)) = \lambda T^i(\vec{x}(t+1)) + U[\vec{x}(t), \vec{\tau}^{i+1}(\vec{x}(t))]$. When $i = 0$, $T^1(\vec{x}(t)) = U[\vec{x}(t), \vec{\tau}^0(\vec{x}(t))] + \lambda T^0(\vec{x}(t+1))$, $T^1(\vec{x}(t)) - T^0(\vec{x}(t)) = U[\vec{x}(t), \vec{\tau}^0(\vec{x}(t))] \geq 0$, we get $T^1(\vec{x}(t)) \geq T^0(\vec{x}(t))$.

Assuming $t = i - 1$, $T^i(\vec{x}(t)) \geq T^{i-1}(\vec{x}(t))$, When $t = i$, $T^{i+1}(\vec{x}(t)) = U[\vec{x}(t), \vec{\xi}^i(\vec{x}(t))] + \lambda T^i \vec{x}(t+1)$ and $T^{i+1}(\vec{x}(t)) - T^i(\vec{x}(t)) = \lambda(U[\vec{x}(t), \vec{\xi}^{i-1}(\vec{x}(t+1))]) \geq 0$. Then $T^{i+1}(\vec{x}(t)) \geq T^i(\vec{x}(t))$. And we can get $K^i(\vec{x}(t)) \leq K^{i+1}(\vec{x}(t))$.

Theorem 3.7 *It is known that* $\{\vec{\kappa}^i(\vec{x}(t))\}$ *is the control sequence corresponding to the minimum cost function* $K^i(\vec{x})$, *which can prove* $\lim_{i\to\infty} K^i(\vec{x}(t)) = K^*(\vec{x}(t))$.

Proof $\{\kappa^i(\vec{x})\}$and $K^i(\vec{x})$ have been given in Lemma 3.2, and the corresponding value function of $\{\kappa^{i,l}(\vec{x})\}$ is $K^{i+1,l}(\vec{x}(t)) = U[\vec{x}(t), \vec{\kappa}^{i,l}(\vec{x}(t))] + \lambda K^{i,l}(\vec{x}(t))$, where l is the length. Obviously, $K^{i+1,l}(\vec{x}(t)) = \sum_{j=0}^{i} \lambda^i U[\vec{x}(t+j), \vec{\kappa}^{i-j,l}(\vec{x}(t+j))]$.

After taking the limit, we can obtain $K^{\infty,l}(\vec{x}(t)) = \lim_{i\to\infty} \sum_{j=0}^{i} \lambda^i U[\vec{x}(t+j), \vec{\kappa}^{i-j,l}(\vec{x}(t+j))]$, and define $K^*(\vec{x}(t)) = \inf_l\{K^{\infty,s}(\vec{x}(t))\}$. Similarly, $\Omega^{\infty+1,s}(\vec{x}(t)) \leq K^{\infty,l}(\vec{x}(t)) \leq D^s$ can be obtained from Lemma 3.5. On the other hand, we get $K^{i+1}(\vec{x}(t)) \leq K^{\infty,s}(\vec{x}(t))$ based on Lemma 3.4. Therefore, it can be concluded that $K^{i+1}(\vec{x}(t)) \leq \Omega^{i+1,l}(\vec{x}(t)) \leq \Omega^{\infty,l}(\vec{x}(t)) \leq D^s$. $K^*(\vec{x}(t)) = \inf_l K^{\infty,l}(\vec{x}(t))$ with the definition of minimum value for the optimal value equation, extracting a control sequence $\{\vec{\kappa}^{i,m}\}$ so that $K^{\infty,m} \leq K^*(\vec{x}(t)) + \epsilon$, and drawing an conclusion that $K^{\infty,m} \leq K^*(\vec{x}(t)) + \epsilon$. Considering $K^{i+1}(\vec{x}(t)) \leq K^{i+1,l}(\vec{x}(t)) \leq K^{\infty,l}(\vec{x}(t)) \leq D^l$ in another way and taking the limit, the formula holds for any i, l, then $\lim_{i\to\infty} K^i(\vec{x}(t)) = \inf_s D^s$.

To guarantee $\lim_{i\to\infty} K^i(\vec{x}(t)) = K^{\infty,g}(\vec{x}(t))$, the control sequence $\{\vec{\kappa}^{i,g}\}$ is necessary, and then we can get $K^{i+1}(\vec{x}(t)) \geq K^*(\vec{x}(t))$. Combining both aspects above, $\lim_{i\to\infty} K^i(\vec{x}(t)) = K^*(\vec{x}(t))$ is obtained.

Theorem 3.8 *For any state variable* $\vec{x}(t)$, *the optimal value equation* $K^i(\vec{x}(t))$ *satisfies the characteristics of the HJB equation.*

$$K^*(\vec{x}(t)) = U[\vec{x}(t), \vec{\kappa}(t)] + \lambda K^*(\vec{x}(t+1)). \tag{3.35}$$

Proof From the proved lemmas and theorems, a series of characteristics about "$K^i(\vec{x}(t))$" are obtained. At this time, it is necessary to verify that characteristics of the HJB equation are satisfied. According to (3.23), there exits $K^*(\vec{x}(t)) = \inf_{\vec{\kappa}(t)}\{U[\vec{x}(t), \vec{\kappa}] + \lambda K^i(\vec{x}(t+1))\}$, meanwhile, according to Theorems 3.6 and 3.7, yield that $K^{i+1}(\vec{x}(t)) = \min_{\vec{\kappa}(t)}\{U[\vec{x}(t), \vec{\kappa}] + \lambda K^i(\vec{x}(t+1))\}$. Then take the mathematical limit, we get $K^*(\vec{x}(t)) \leq \inf_{\vec{\kappa}(t)} U[\vec{x}(t), \vec{\kappa}] + \lambda K^i(\vec{x}(t+1)$ for the randomness of $\{\vec{u}(t)\}$).

From the other side, we have $K^{i+1}(\vec{x}(t)) \geq \inf_{\vec{\kappa}(t)} U[\vec{x}(t), \vec{\kappa}] + \lambda K^{i-1}(\vec{x}(t+1)$, take the limit again, then yield that $K^*(\vec{x}(t)) \geq \inf_{i} U[\vec{x}(t), \vec{\kappa}] + \lambda K^i(\vec{x}(t+1))$. As to the analysis above, we can get a final conclusion.

$$K^*(\vec{x}(t)) = U[\vec{x}(t), \vec{\kappa}(t)] + \lambda K^*(\vec{x}(t+1)). \tag{3.36}$$

All content is verified.

Remark 3.9 The control sequence $\{\vec{\kappa^i}(\vec{x}(t))\}$ is a monotonic non-decreasing sequence corresponding to the minimum value function $K^i(\vec{x})$, which tend to be $K^*(\vec{x}(t))$ eventually, satisfying the characteristics of the HJB functions as [31].

3.4 Simulation and Numerical Experiments

In this section, we consider the mechanism model of tumor cell growth combined with immunotherapy, chemotherapy and combination treatments proposed as experimental validation. Firstly, The affine system model is constructed with chemotherapy drugs and immunoagents as control inputs and the account involved of tumor cells as state variables. Secondly, according to the affine model obtained by fitting, we developed the cost function of treatment loss with the clinical treatment requirements. Finally, the optimal treatment plan for a patient with a basic condition is given after calculation by the algorithm.

3.4.1 An Affine Model of Tumor Cell Growth

According to clinical medical statistics and literature [4], the specific parameters of the mechanism model are given as Table 3.3.

At this point, when we give the initial count of tumor cell population and immune cells in a patient and follow a certain chemotherapy and immunotherapy regimen, we can get the following four curves on tumor cells and immune cell population as shown in Figs. 3.1 and 3.2. It is obviously that state variable $Tu(t)$ denoted the population of tumor cells tend to be stable in Fig. 3.2, similarly, for $Im(t)$ in Fig. 3.1.

When the fitted affine system is carried out according to the data obtained from the mechanism model, $Dr_{Che}(t)$ and $Dr_{Im}(t)$ are selected as two control inputs and $Im(t)$ as state variables. Within the allowable error range, the obtained fitting relation is shown as the following form,

Table 3.3 Concentration variation on immune cells, tumor cells, chemotherapeutic drug and immunoagents

Parameter	Estimated value	Units	Parameter	Estimated value	Units
C_1	0.00431	day^{-1}	C_2	1.02×10^{-9}	cell^{-1}
C_3	0.33	$\text{cell} \cdot \text{day}^{-1}$	$C_{Im,Tu}$	6.41×10^{-11}	cell^{-1}
$C_{Che,Tu}$	0.08	day^{-1}	$C_{Im,d}$	0.204	day^{-1}
$C_{Tu,Im}$	3.42×10^{-6}	cell^{-1}	$C_{Che,Im}$	2×10^{-11}	day^{-1}
α_1	0.0125	day^{-1}	α_2	0.125	day^{-1}
β_1	2.02×10^7	cell	β_2	2×10^7	cell
γ_1	0.1	day^{-1}	γ_2	1	day^{-1}

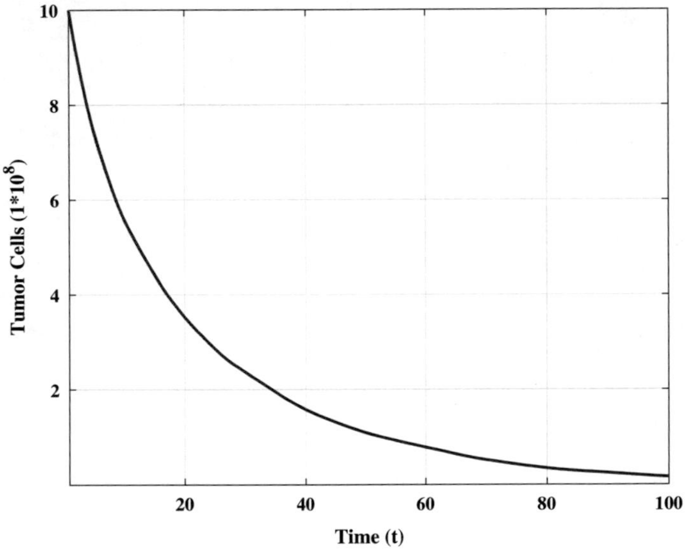

Fig. 3.1 The curves of tumor cells

$$x(t+1) = f(x(t)) - [g_1(x(t)), g_2(x(t))][u_1(t), u_2(t)]^T. \tag{3.37}$$

$$f(x) = x + 0.00431x(1 - 1.02 \times 10^{-9}x). \tag{3.38}$$

$$g_1(x) = exp(8.15 \times 10^{-6}[log(x)]^{6.131} + 3.482). \tag{3.39}$$

$$g_2(x) = exp(0.05639[log(x)]^{2.093} + 2.492). \tag{3.40}$$

The curves before and after fitting are compared as Fig. 3.3, which meets the requirements of fitting precision, which guarantees accuracy of the data traced back to the original source.

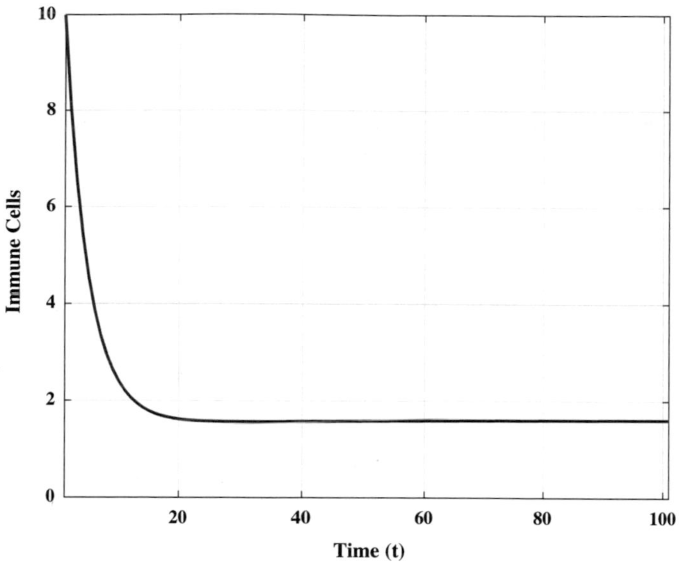

Fig. 3.2 The curves of immune cells

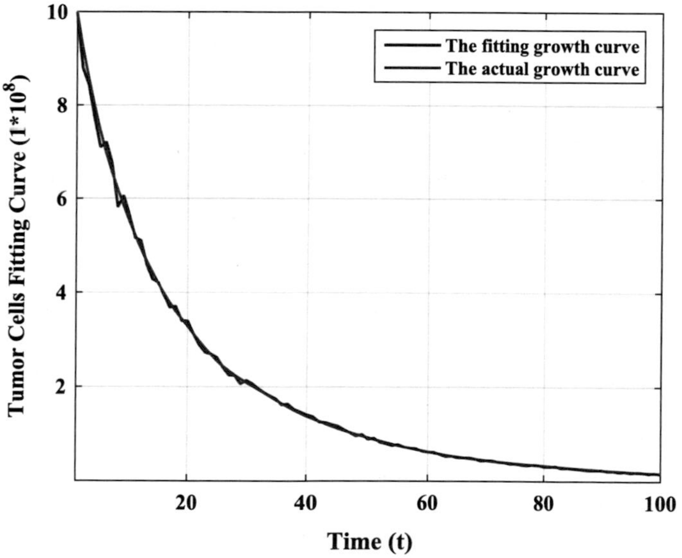

Fig. 3.3 The curves of immunoagents drug concentrations in the bloodstream

3.4.2 The Treatment Loss Cost Function

The form of the cost function proposed in the third part as (3.17). Unlike the theoretical mechanism model analysis, and combined with clinical requirements, it is necessary to limit the single injection of drugs to no more than 0.05. Therefore,

$$\bar{U}_1 = 0.05, \bar{U}_2 = 0.05. \tag{3.41}$$

To avoid the optimal solution in the infinite time dimension, we choose the discount factor $\lambda = 0.95$. Finally, the specifically obtained cost function as follows:

$$K(x) = \sum_{t=t_0}^{\infty} 0.95^t \{2.784 \times 10^{-5} x(t)^2 + \int_0^{u_1(t)} 50tanh^{-1} (0.05^{-1}s)ds + \int_0^{u_2(t)} 850tanh^{-1}(0.05^{-1}s)ds\}. \tag{3.42}$$

3.4.3 The Optimal Solution of the Treatment

According to the previous two subsections, we have completed the transformation from the mathematical mechanism model to the solvable affine model, and determined the specific value of the cost function according to the clinical requirements. The optimal treatment strategy is acquired through the proposed algorithm and make a comparison to prove the effectiveness and feasibility. The cost function is designed to minimize the tumor cells, meanwhile, there exit minimum dose chemotherapy drugs and immunoagents.

In the following three figures (Figs. 3.4, 3.5 and 3.6), the blue curve represents the changes of tumor cells and the changes of a single dose in patients under the normal treatment regimen. In contrast, the red curve represents the optimal treatment regimen's effect calculated by the nonzero-sum game-based ADP algorithm.

As shown in Fig. 3.4, there are originally many cancer cells in the body. The two curves are close to the upper limit, with drugs and dual function of the immune system, a substantial reduction in the number of cancer cells. The amount of drug injection therapy hasn't changed greatly during the process from beginning to end. Even in the closing stage, cancer cells decreased significantly, there are still specific doses, and we solve the treatment dose is substantially less than the former.

Correspondingly, as shown in Fig. 3.5 that the changing trend of the injection dose of immunoagents on the two curves is close to the changing direction of chemotherapy drugs. The optimized treatment is slightly more than the traditional treatment plan when more cancer cells are in the initial stage, but it will not last for a long time. When the number of cancer cells is relatively large, the primary or indirect target of these two drugs is cancer cells; then, in the late stage of treatment, the number of

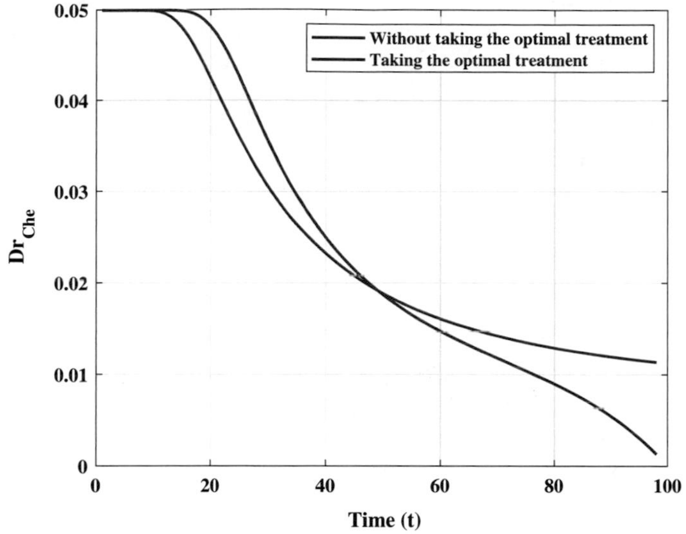

Fig. 3.4 The injection dose curve of chemotherapy drugs under two kinds of treatment

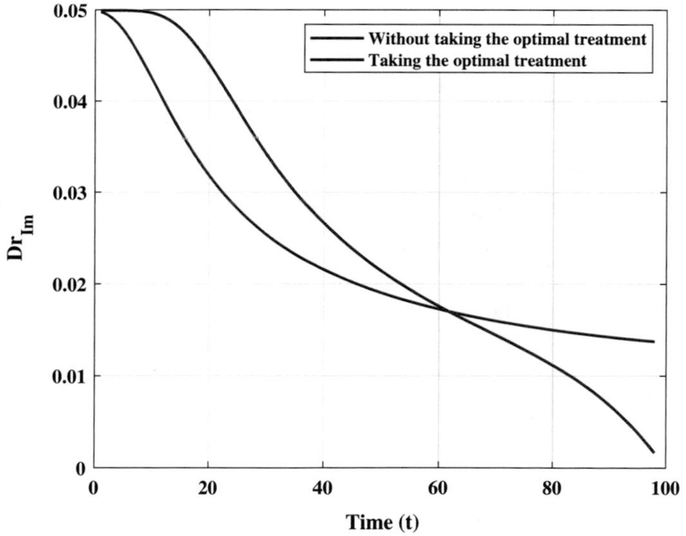

Fig. 3.5 The injection dose curve of immunologic agents under two kinds of treatment

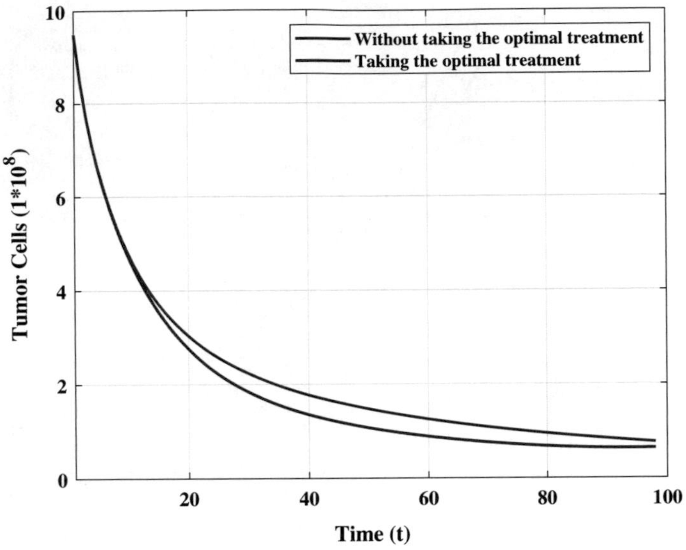

Fig. 3.6 The curves of tumor cells under two kinds of treatment

cancer cells is significantly reduced. If the chemotherapy drugs are put in according to the normal treatment, the normal cells will suffer a lot of erosion, which has a more significant impact on the body. However, the optimized drug dose has been dramatically reduced, and the normal cells have been less affected.

As shown in Fig. 3.6, control effect of the two treatment schemes on the number of tumor cells enjoy resemblance to that in the initial stage. Still, at the final stage, the algorithm optimized by ADP not only significantly reduces the count of tumor cell population, combined with Figs. 3.4 and 3.5, but also minimize the injection amount of the two drugs, which shows the effectiveness of our treatment scheme.

Remark 3.10 The optimal regulation strategy for the immune system enjoys advantage of decreasing of tumor cells, what is more, clinical treatment benefits from typical minimization of chemotherapy drugs and immunoagents.

3.5 Conclusion

Nonzero-sum games-based adaptive dynamic programming has been proposed acquiring the optimum through affecting the growth of tumor and immune cells, providing guidance for clinical practice through adjusting the administered doses of chemotherapy drugs and immunotherapy drugs. Obtained results have shown that the immune system can decrease the tumor cells, meanwhile, minimizing of chemotherapy drugs and immunoagents through optimal control behavior. Simulation examples have presented availability and effectiveness of the research methodology. The future research will focus on solving the optimal mixed treatment strategy taking account of complex immunotherapy system including immune cell subsets and cytokines, considering the switched control policies in according with hybrid therapy.

References

1. Wang J, Huang M, Chen S, Luo Y, Shen S, Du X (2021) Nanomedicine-mediated ubiquitination inhibition boosts antitumor immune response via activation of dendritic cells. Nano Res 14:3900–3906
2. Chen C, Li A, Sun P, Xu J, Du W, Zhang J, ..., Jiang X (2020) Efficiently restoring the tumoricidal immunity against resistant malignancies via an immune nanomodulator. J Control Rel 324(10):574–585
3. Zhang T, Chen CP, Chen L, Xu X, Hu B (2018) Design of highly nonlinear substitution boxes based on I-Ching operators. IEEE Trans Cybernet 48(12):3349–3358
4. de Pillis LG, Gu W, Radunskaya AE (2006) Mixed immunotherapy and chemotherapy of tumors: Modeling, applications and biological interpretations. J Theor Biol 238(4):841–862
5. Ogunmadeji B, Yusuf T (2018) Optimal control strategy for improved cancer biochemotherapy outcome. Int J Sci Eng Res 9(12):583–600
6. Chen CP, Wen GX, Liu YJ, Liu Z (2016) Observer-based adaptive backstepping consensus tracking control for high-order nonlinear semi-strict-feedback multiagent systems. IEEE Trans Cybernet 46(7):1591–1601
7. Wang D, Ha M, Qiao J (2020) Self-learning optimal regulation for discrete-time nonlinear systems under event-driven formulation. IEEE Trans Autom Control 65(3):1272–1279
8. Zhang T, Su G, Qing C, Xu X, Cai B, Xing X (2021) Hierarchical lifelong learning by sharing representations and integrating hypothesis. IEEE Trans Syst Man Cybernet: Syst 51(2):1004–1014
9. Huang H, Zhang T, Yang C, Chen CP (2020) Motor learning and generalization using broad learning adaptive neural control. IEEE Trans Ind Electron 67(10):8608–8617
10. Zhang H, Cui L, Luo Y (2013) Near-optimal control for nonzero-sum differential games of continuous-time nonlinear systems using single-network ADP. IEEE Trans Cybernet 43(1):206–216
11. Li T, Yang D, Xie X, Zhang H (2022) Event-triggered control of nonlinear discrete-time system with unknown dynamics based on HDP(λ). IEEE Trans Cybernet 52(7):6046–6058
12. Zhao B, Liu D (2020) Event-triggered decentralized tracking control of modular reconfigurable robots through adaptive dynamic programming. IEEE Trans Ind Electron 67(4):3054–3064
13. Liang H, Liu G, Zhang H, Huang T (2021) Neural-network-based event-triggered adaptive control of nonaffine nonlinear multiagent systems with dynamic uncertainties. IEEE Trans Neural Netw Learn Syst 32(5):2239–2250
14. Sun J, Zhang H, Wang Y, Sun S (2022) Fault-tolerant control for stochastic switched IT2 fuzzy uncertain time-delayed nonlinear systems. IEEE Trans Cybernet 52(2):1335–1346
15. Sun J, Zhang H, Wang Y, Shi Z (2022) Dissipativity-based fault-tolerant control for stochastic switched systems with time-varying delay and uncertainties. IEEE Trans Cybernet 52(10):10683–10694
16. Doyle JC, Glover K, Khargonekar PP, Francis BA (1989) State-space solutions to standard H_2 and H_∞ control problems. IEEE Trans Autom Control 34(8):831–847
17. Davari M, Gao W, Jiang ZP, Lewis FL (2021) An optimal primary frequency control based on adaptive dynamic programming for islanded modernized microgrids. IEEE Trans Autom Sci Eng 18(3):1109–1121
18. Yang D, Li T, Xie X, Zhang H (2020) Event-triggered integral sliding-mode control for nonlinear constrained-input systems with disturbances via adaptive dynamic programming. IEEE Trans Syst Man Cybernet: Syst 50(11):4086–4096

19. Zhao B, Liu D, Luo C (2020) Reinforcement learning-based optimal stabilization for unknown nonlinear systems subject to inputs with uncertain constraints. IEEE Trans Neural Netw Learn Syst 31(10):4330–4340
20. Yang Y, Ding D-W, Xiong H, Yin Y, Wunsch DC (2020) Online barrier-actor-critic learning for H_∞ control with full-state constraints and input saturation. J Frankl Inst 357(6):3316–3344
21. Zhong X, He H, Wang D, Ni Z (2018) Model-free adaptive control for unknown nonlinear zero-sum differential game. IEEE Trans Cybernet 48(5):1633–1646
22. Luo B, Yang Y, Liu D, Wu H (2020) Event-triggered optimal control with performance guarantees using adaptive dynamic programming. IEEE Trans Neural Netw Learn Syst 31(1):76–88
23. Yang Y, Gao W, Modares H, Xu CZ (2022) Robust actor-critic learning for continuous-time nonlinear systems with unmodeled dynamics. IEEE Trans Fuzzy Syst 30(6):2101–2112
24. Starr AW, Ho YC (1969) Nonzero-sum differential games. J Optim Theory Appl 3(3):184–206
25. Zhang H, Wei Q, Liu D (2011) An iterative adaptive dynamic programming method for solving a class of nonlinear zero-sum differential games. Automatica 47(1):207–214
26. Zhu Y, Zhao D (2022) Online minimax Q network learning for two-player zero-sum Markov games. IEEE Trans Neural Netw Learn Syst 33(3):1228–1241
27. Zhong X, He H (2017) An event-triggered ADP control approach for continuous-time system with unknown internal states. IEEE Trans Cybernet 47(3):683–694
28. Wei Q, Li H, Yang X, He H (2021) Continuous-time distributed policy iteration for multicontroller nonlinear systems. IEEE Trans Cybernet 51(5):2372–2383
29. Wei Q, Song R, Liao Z, Li B, Lewis FL (2020) Discrete-time impulsive adaptive dynamic programming. IEEE Trans Cybernet 50(10):4293–4306
30. Zhang H, Qin C, Jiang B, Luo Y (2014) Online adaptive policy learning algorithm for H_∞ state feedback control of unknown affine nonlinear discrete-time systems. IEEE Trans Cybernet 44(12):2706–2718
31. Yang Y, Vamvoudakis KG, Modares H, Yin Y, Wunsch DC (2021) Hamiltonian-driven hybrid adaptive dynamic programming. IEEE Trans Systems Man Cybernet: Syst 51(10):6423–6434

Chapter 4
Evolutionary Dynamics Optimal Research-Oriented Tumor Immunity Architecture

4.1 Introduction

Interaction between cancer cells, surrounding stromal cells and immune cells through autonomous and non-autonomous signaling can influence survival competition. Therefore, it is very critical for evolutionary and ecological dynamics mechanistic understanding of tumor progression [1]. It is assumed that evolution causes traits to change continuously over time even if the ecological dynamics are constantly changing. More broadly, imagine an evolutionarily stable state that is a trajectory of phenotypic states-an evolutionarily stable trait attractor. This can be used in scenarios where there is sufficient variation to facilitate rapid evolution, or where the state involves a plastic response to environmental conditions, eventually constituting evolutionary stability. Simultaneously, Natural killer (NK) cells as one of the players in the game attack many tumour cell lines, which is critical in anti-tumour immunity [2], however, the interaction between NK cells and tumour targets is poorly. To overcome drug resistance, anti-tumor immunotherapy gradually replaces the traditional treatment strategy [3]. The interaction between specialized cancer cell populations and immune cells has become a special evolutionary dynamics phenomenon in the process of tumor immunity growth architecture. The goal of optimization is to minimize administration dosage and reduce negative effects.

The dynamic perception or learning process is realized through interactions between cells and organism architecture, accomplishing observing their responses and learning optimal control strategy ultimately of Markov decision. It is required to seek an optimal control scheme such that the desired dosage of administration can be tracked and the optimal performance of minimize chemotherapeutic drugs and immunological agents can be achieved. Thus, reinforcement learning is urgently needed for optimal research-oriented tumor immunity architecture. The classical policy iteration and value iteration frameworks are never out of date, and the new min-Hamilton function [4] and the low-gain parameter ADP-Bellman equation for global stabilization are thriving [5].

J. Sun et al., *Adaptive Dynamic Programming*,
https://doi.org/10.1007/978-981-99-5929-7_4

The interaction between cells is highly nonlinear and coupled. When the computational conditions allow, whether it is the adaptive algorithm design based on policy iteration, or the adaptive hierarchical neural network algorithm [6], which can easily solve the coupled fractional order chaotic synchronization problem. All inspire us in solving the optimal solution of the HJB equation of the idea. Once computing conditions are not available, model-free is the best idea. The iteration-ADP algorithm is developed into iteration-NDP algorithm, which does not require an accurate system model [7], but only requires observable system data, which can reduce the cost and optimize the control action in the process of error backpropagation [8]. The emergence of Q-learning, from containing three classes four networks to interleaving double iteration, and then to the critical Q-learning [9] of a single class one network, effectively improves the utilization of resources, and the problem of insufficient exploration no longer exists. The interaction between cells coincides with multiple agents, and the attack of tumor cells on normal cells may cause abnormal reactions, and the neural net-based attack detection and estimation scheme designed by [10] can easily capture such anomalies. Cells cannot proliferate without limit. When solving the optimal solution of the constrained auxiliary subsystem, based on the framework of ADP, the idea of pi iteration is continued, and a strong convergence synchronous iterative optimization strategy [11] is given.

The difficult-to-decouple leaderer-follower behavior of vehicle-vehicle communication [12], human-vehicle interaction, and mutual quality of everyone can be easily solved with off-PI [13]. Switching system [14], T-S fuzzy, nash equilibrium, zero-sum game [15], let each agent deal with a low-dimensional state and local pattern, reduce conservatism, can easily obtain the minimum local cost [16]. Influenced by the improved exploration feature, the parallel A-C asynchronous gradient sharing mechanism can realize the parallel optimization operation of diversified agents in a short time [17]. Affected by the time difference error, integral reinforcement learning can obtain the estimated control strategy by updating the critic weight [18, 19]. In order to obtain a better stabilizing adaptive control scheme, it is necessary to give an appropriate robust control scheme for the control system [20]. Reference [21] summarizes the recent outstanding progress in the continuous nonlinear control system of the controller that combines adaptability and robustness. The reliability and effectiveness of the actual power system and some large machinery and heavy machinery devices with these two designs considered are also demonstrated. The theory integrates ecological and evolutionary dynamics blending ecological mathematical model evolutionary game theory [22]. Then evolutionarily stable strategies will be investigated to seamless integration of both sides [23]. Solvable dynamic equations can be used to explore optimal control objectives, however, what followed is a disaster of dimensions.

To overcome it, dual-heuristic dynamic programming is proposed for the nonlinear affine evolutionary dynamic dated from ADP considering the actual constraints. By introducing a discounted performance index, the optimal regulation problem of the infinite dimensional problem is reformulated into a finite dimensional. Different from previous value iterations which requires a strategy for initially stable the system. ADP is conformed to the optimal formation control by the establishment of performance index function [24]. The affine mathematical model is firstly introduced to twinborn the real scenario [25]. The optimal control is transformed into pursuing solution of HJB, and the convergence is proved. ADP involves learners giving rise to learning strategy, and the author studied a competitive learning system setting with cancer cell populations and immune cells, aiming at minimizing the dose administered.

4.2 Pre-knowledge

Consider a classical discrete-time nonlinear affine system,

$$x(t+1) = f(x(t)) + g(x(t))u(t) \tag{4.1}$$

where the state variable $x(t) \in \mathcal{R}^n$, the control variable $u(t) \in \mathcal{R}^m$, and $f(\cdot) \in \mathcal{R}^n$, $g(\cdot) \in \mathcal{R}^{n\times m}$ can be stabilized on a compact set $\Omega \in \mathcal{R}^n$, and $f(0) = 0\, g(0) = 0$. Colloquially, the optimal control problem of (4.1) is equivalent to obtaining $u^*(t) = u(x(t))$(the optimal control law) that minimizes the proposed infinite-horizon performance index:

$$\mathrm{J}(x(t)) = \sum_{t=0}^{\infty} \mathrm{K}(x(t), u(t)). \tag{4.2}$$

$\mathrm{K}(x(t), u(t))$ is the cost function, $\mathrm{K}(x, u) \geq 0\ \forall x, u$. Basically, the cost function $\mathrm{K}(\cdot)$ is given a quadratic form

$$\mathrm{K}(x(t), u(t)) = x^T(t)\mathrm{P}x(t) + u^T(t)\mathrm{Q}u(t) \tag{4.3}$$

$\mathrm{P}, \mathrm{Q} > 0$ are all positive definite matrices.

The optimal control problem of (4.2) can be converted to solve the HJB equation. According to the Bellman optimal principle, the optimal value function should obey the following[9]:

$$\mathrm{J}^*(x(t)) = \min_{u(t)} \left\{ x^T(t)\mathrm{P}x(t) + u^T(t)\mathrm{Q}u(t) + \mathrm{J}^*(x(t+1)) \right\} \tag{4.4}$$

By minimizing the right side of the (4.4) to solve the optimal control law, get the optimal value function $\mathrm{J}^*(x(t))$. For necessity, one can take the partial derivative of the right-hand side of (4.4) with respect to $u(t)$ to obtain u^*. Hence,

$$u^*(t) = -\frac{\mathrm{Q}^{-1}}{2} \left[g(x(t))\right]^T \frac{\partial \mathrm{J}^*(x(t+1))}{\partial x(t+1)} \tag{4.5}$$

Take (4.5) into (4.4), it can be obtained that

$$\begin{aligned} J^*(x(t)) =& x^T(t)Px(t) + \frac{1}{4}\Big[\frac{\partial J^*(x(t+1))}{\partial x(t+1)}\Big]^T g(x(t))Q^{-1} \\ & \cdot g^T(x(t))\Big[\frac{\partial J^*(x(t+1))}{\partial x(t+1)}\Big] + J^*(x(t+1)). \end{aligned} \tag{4.6}$$

By the on (4.6), it is almost impossible to obtain an analytical solution for $u^*(t)$. Impossible in the current moment t can know the next moment $J^*(x(t+1))$. To overcome this dilemma, the approximate optimal solution of HJB equation can be studied. In the fourth part of this chapter, the derivation of IDHP algorithm is introduced to solve this kind of optimal control problem [26, 27].

4.3 Modeling of Mixed Immunotherapy and Chemotherapy for Tumor Cell

In this part, a mathematical model is constructed from the natural growth of a single type of tumor cells, the gradual increase of the interaction between various immune cells and tumor cells in vivo, and the influence of external application of chemotherapy drugs and immune agents on the population of tumor cells [22, 28, 29].

First, define the acronyms of various cells:

- $\mathcal{T}_u(t)$: Tumor cell population in the vivo.
- $\mathcal{n}_{\mathcal{K}}(t)$: NK cells are derived from bone marrow lymphoid stem cells.
- $\mathcal{C}_{\mathcal{T}}(t)$: Cytotoxic T lymphocytes (CTL), a subdivision of leukocytes, are specific T cells that secrete various cytokines and participate in immune function.
- $\mathcal{C}_{\mathcal{L}}(t)$: Number of circulating lymphocytes (or leukocytes).
- $\mathcal{C}h_{\mathfrak{dr}}(t)$: Chemotherapeutic drug concentration in the blood.
- $\mathcal{I}m_{\mathfrak{dr}}(t)$: Immunotherapy drug concentration in the blood.

For the convenience of writing, the following subsections do not specify the time, and the default is t, lowercase letters "$\mathfrak{a}$, $\mathfrak{b}$, $\mathfrak{c}_1$, $\mathfrak{c}_2$, $\mathfrak{e}$, $\mathfrak{f}$, $\mathfrak{g}$, $\mathfrak{h}_1$, $\mathfrak{h}_2$, $\mathfrak{i}$, $\mathfrak{j}$, $\mathfrak{l}$, $\mathfrak{m}$, $\mathfrak{n}_1$, $\mathfrak{n}_2$, $\mathfrak{p}_1$, $\mathfrak{p}_2$, $\mathfrak{q}_1$, $\mathfrak{q}_2$, $\mathfrak{r}$, $\mathfrak{s}$, $\mathfrak{u}$" all represent fixed real numbers; Uppercase letters "G, K, O, R, I" represent different categories of gain items, which depend on time t; $\mathcal{L}_{(\cdot)}$ is a constant that depends on the cell type; and $\mathbf{e}^{(\cdot)}$ stands for exponential functions.

4.3.1 *The Natural Growth of Cells*

According to the [2, 22], the increase of tumor cells follows a natural growth curve, $\mathcal{G}_{\mathcal{T}_u} = \mathfrak{a}\mathcal{T}_u(1 - \mathfrak{b}\mathcal{T}_u)$ ($\mathcal{G}_{(\cdot)}$ represents the natural growth tescr operator of all types of cells). Natural killer cells [22] are assumed to be produced at a constant rate and to be influenced by circulating lymphocytes throughout the production cycle (since circulating lymphocytes represent the overall level of immune health), and thus,$\mathcal{G}_{\mathcal{N}_{\mathcal{K}}} = \mathfrak{c}_1 \mathcal{C}_{\mathcal{L}} - \mathfrak{c}_2 \mathcal{N}_{\mathcal{K}}$. In the absence of tumor cells, Cytotoxic T lymphocytes are assumed to be absent and cell growth of $\mathcal{C}_{\mathcal{T}}(t)$ cells is only affected by natural mortality, $\mathcal{G}_{\mathcal{C}_{\mathcal{T}}} = -\mathfrak{e}\mathcal{C}_{\mathcal{T}}$. Circulating lymphocytes are also produced at a constant rate during their lifetime, $\mathcal{G}_{\mathcal{C}_{\mathcal{L}}} = \mathfrak{f} - \mathfrak{g}\mathcal{C}_{\mathcal{L}}$. It is set that when the body is injected with chemotherapy drugs or immune agents, it will show exponential decay, $\mathcal{G}_{\mathcal{C}h_{\mathfrak{dr}}} = -\mathbf{e}^{-\gamma_\alpha}\mathcal{C}h_{\mathfrak{dr}}$, $\mathcal{G}_{\mathcal{I}m_{\mathfrak{dr}}} = -\mathbf{e}^{-\gamma_\beta}\mathcal{I}m_{\mathfrak{dr}}$.

4.3.2 *Intercellular Conditioning*

When the above cells exist at the same time, there will be a negative interaction between the two populations, partly due to the competition for growth space and nutrients, and this indirect effect. The other part is the direct resistance of cell populations to each other [22]:

$$\mathcal{K}_{\mathcal{T}_u} = -\mathfrak{j}\mathcal{N}_{\mathcal{K}}\mathcal{T}_u \quad \mathcal{K}_{\mathcal{C}_{\mathcal{T}}} = \mathfrak{h}_1 \cdot \frac{(\mathcal{C}_{\mathcal{T}}/\mathcal{T}_u)^{\mathfrak{i}}}{\mathfrak{h}_2(\mathcal{C}_{\mathcal{T}}/\mathcal{T}_u)^{\mathfrak{i}}} \cdot \mathcal{T}_u$$

And just to simplify the writing, let's write $\mathfrak{O}$ for a particular term, and notice that $\mathfrak{O} = \mathfrak{O}(t)$, which is related to $\mathcal{C}_{\mathcal{T}}(t)$, $\mathcal{T}_u(t)$.

$$\mathfrak{O} = \mathfrak{h}_1 \cdot \frac{(\mathcal{C}_{\mathcal{T}}/\mathcal{T}_u)^{\mathfrak{i}}}{\mathfrak{h}_2(\mathcal{C}_{\mathcal{T}}/\mathcal{T}_u)^{\mathfrak{i}}} \quad \mathcal{K}_{\mathcal{C}_{\mathcal{T}}} = \mathfrak{O} \cdot \mathcal{T}_u \tag{4.7}$$

NK cells have the function of recruitment, which is to design sequential application methods of cell cycle non-specific drugs and cell cycle specific drugs, recruit more cells at specific stages into the proliferation cycle, so as to increase the number of tumor cells killed [29–31].

$$\mathcal{R}_{\mathcal{N}_{\mathfrak{k}}} = \frac{\mathfrak{l} \cdot {\mathcal{T}_u}^2}{\mathfrak{m}{\mathcal{T}_u}^2}\mathcal{N}_{\mathfrak{k}}; \quad \mathcal{R}_{\mathcal{C}_{\mathcal{T}}}(\mathcal{T}_u, \mathcal{C}_{\mathcal{T}}) = \mathfrak{p}_1 \frac{\mathfrak{O}^2{\mathcal{T}_u}^2}{\mathfrak{q}_1 + \mathfrak{O}^2{\mathcal{T}_u}^2}\mathcal{C}_{\mathcal{T}}$$

$\mathcal{C}_{\mathcal{T}}$ cells have a similar recruitment effect [32]. It is directly proportional to the number of cells killed by NK cell lysis of tumor cells, $\mathcal{R}_{\mathcal{C}_{\mathcal{T}}}(\mathcal{N}_{\mathfrak{k}}, \mathcal{T}_u) = \mathfrak{n}_1\mathcal{N}_{\mathfrak{k}}\mathcal{T}_u$. Also, the presence of tumor cells stimulates the immune system to secrete more cells, $\mathcal{R}_{\mathcal{C}_{\mathcal{T}}}(\mathcal{C}_{\mathcal{L}}, \mathcal{T}_u) = \mathfrak{n}_2\mathcal{C}_{\mathcal{L}}\mathcal{T}_u$. In the immune function, NK cells or CD cells may have to undergo multiple contact with tumor cells, and then inactivate [29, 33–35].

$$\mathcal{I}_{ac,\mathcal{N}_{\mathfrak{k}}} = -\mathfrak{p}_2\mathcal{T}_u\mathcal{N}_{\mathfrak{k}} \quad \mathcal{I}_{ac,\mathcal{C}_{\mathcal{T}}} = -\mathfrak{q}_2\mathcal{C}_{\mathcal{T}}\mathcal{T}_u \quad \mathcal{I}_{\mathcal{C}_{\mathcal{L}},\mathcal{C}_{\mathcal{T}}} = -\mathfrak{r}\mathcal{N}_{\mathfrak{k}}(\mathcal{C}_{\mathcal{T}})^2$$

4.3.3 Drug Intervention

All kinds of cell populations in this model contain the action tescr of chemotherapy drugs, and the killing effect of chemotherapy drugs is not always effective. At low drug concentration, the killing rate increases almost linearly, while at high drug concentration, the killing rate tends to be stable. Saturation type is used to describe them in the model [36], $1-\mathbf{e}^{\mathcal{C}h_{dz}(t)}$.

$$\mathcal{D}_r^{\mathcal{C}h}(\cdot)=\mathcal{L}_{(\cdot)}(1-\mathbf{e}^{\mathcal{C}h_{dz}(t)})(\cdot)$$

$(\cdot)=\mathcal{T}_u,\mathcal{C}_{\mathcal{T}},\mathcal{C}_{\mathcal{L}},\mathcal{N}_{\mathfrak{k}}$.

$\mathcal{L}_{(\cdot)}$ represents the interaction coefficient between corresponding cells and tumor cells. It also includes immunotherapy, whose impact on immune system efficacy can be mathematically described by the Michaelis-Menten interaction, $\mathfrak{s}$, $\mathfrak{u}$ are the constant [30].

$$\mathcal{D}_r^{\mathcal{I}m}(\mathcal{C}_{\mathcal{T}},\mathcal{I}m_{dz})=\mathfrak{u}\frac{\mathcal{I}m_{dz}\mathcal{C}_{\mathcal{T}}}{s+\mathcal{I}m_{dz}}$$

Chemotherapy and immunotherapy drugs are injected in a certain period of time, and denote by $\mathcal{V}_{Che}(t)$ and $\mathcal{V}_{Im}(t)$ the amount of chemotherapy drug injection and the amount of immunotherapy drug injection, respectively.

4.3.4 Mixed Growth Model of Cell Population

Combined with the above contents, the total cell population growth model can be obtained:

$$\mathcal{I}m_{dz}(t+1)=(1-\mathbf{e}^{-\gamma_\beta})\mathcal{I}m_{dz}(t)+\mathcal{V}_{Im}(t) \tag{4.8a}$$

$$\mathcal{C}he_{dz}(t+1)=(1-\mathbf{e}^{-\gamma_\alpha})\mathcal{C}h_{dz}(t)+\mathcal{V}_{Che}(t) \tag{4.8b}$$

$$\mathcal{C}_{\mathcal{L}}(t+1)=\mathfrak{f}-\mathcal{L}_{\mathcal{C}_{\mathcal{L}}}+(1-\mathfrak{g})\mathcal{C}_{\mathcal{L}}(t)-\mathcal{L}_{\mathcal{C}_{\mathcal{L}}}\mathbf{e}^{\mathcal{C}h_{dz}(t)} \tag{4.8c}$$

$$\begin{aligned}\mathcal{T}_u(t+1)&=(1+\mathfrak{a}-\mathcal{L}_{\mathcal{T}_u})\mathcal{T}_u(t)-\mathfrak{b}\mathcal{T}_u{}^2(t)\\&+\mathcal{T}_u(t)\Big[\mathbf{e}^{\mathcal{C}h_{dz}(t)}-\mathfrak{j}\mathcal{N}_{\mathfrak{k}}(t)-\mathfrak{O}(t)\Big]\end{aligned} \tag{4.8d}$$

$$\begin{aligned}\mathcal{C}_{\mathcal{T}}(t+1)&=(1-\mathfrak{e}-\mathcal{L}_{\mathcal{C}_{\mathcal{T}}})\mathcal{C}_{\mathcal{T}}(t)+[\mathfrak{n}_1\mathcal{N}_{\mathfrak{k}}(t)-\mathfrak{q}_2\mathcal{C}_{\mathcal{T}}(t)\\&+\mathfrak{n}_2\mathcal{C}_{\mathcal{L}}(t)]\cdot\mathcal{T}_u(t)-\mathfrak{r}\mathcal{N}_{\mathfrak{k}}(t)\mathcal{C}_{\mathcal{T}}{}^2(t)+\mathcal{L}_{\mathcal{C}_{\mathcal{T}}}\mathcal{C}_{\mathcal{T}}(t)\\&\cdot\mathbf{e}^{\mathcal{C}h_{dz}(t)}+\mathcal{C}_{\mathcal{T}}(t)\Big[\frac{\mathfrak{u}\mathcal{I}m_{dz}(t)}{s+\mathcal{I}m_{dz}(t)}+\frac{\mathfrak{p}_1\mathfrak{O}^2(t)\mathcal{T}_u{}^2(t)}{\mathfrak{q}_1+\mathfrak{O}^2(t)\mathcal{T}_u{}^2(t)}\Big]\end{aligned} \tag{4.8e}$$

$$\begin{aligned}\mathcal{N}_{\mathfrak{k}}(t+1)&=-\mathcal{L}_{\mathcal{N}_{\mathfrak{k}}}+(1-\mathfrak{c}_2)\mathcal{N}_{\mathfrak{k}}(t)+\frac{\mathfrak{l}\cdot\mathcal{T}_u{}^2(t)}{\mathfrak{m}+\mathcal{T}_u{}^2(t)}\mathcal{N}_{\mathfrak{k}}(t)\\&+\Big[\mathcal{L}_{\mathcal{N}_{\mathfrak{k}}}\mathbf{e}^{\mathcal{C}h_{dz}(t)}-\mathfrak{p}_2\mathcal{T}_u(t)\Big]\mathcal{N}_{\mathfrak{k}}(t)+\mathfrak{c}_1\mathcal{C}_{\mathcal{L}}(t)\end{aligned} \tag{4.8f}$$

4.4 Iterative-Dual Heuristic Dynamic Programming Algorithm for Mixed Treatment

The optimal control problem has been transformed into solving the HJB equation (4.4). In this part, a constrained iterative dual heuristic dynamic programming algorithm based on mixed treatment is given. The algorithm is derived from adaptive dynamic programming [26]. This part mainly three parts research content are presented as working mechanism of ADP algorithm, structure of constrained iterative dual-heuristic dynamic programming algorithm and proof of convergence on I-DHP algorithm.

4.4.1 *Working Mechanism of ADP Algorithm*

Generally speaking, for unconstrained control problems, the performance functional (4.3) is usually chosen as the quadratic form. In this chapter, considering the actual constraints, is transformed into solving a bounded control problem, adopted a non-quadratic functional as follows:

$$\mathrm{Y}(t) = x^{T}(t)\mathrm{P}x(t) + 2\int_{0}^{u(t)} \tanh^{-T}(\overline{\mathcal{U}}^{-1}s)\overline{\mathcal{U}}\mathrm{Q}ds$$

It is convenient for mathematical calculation avoiding the loop or unlimited create unlimited returns markov decision process. In the loop or unlimited markov process which will constantly get reward again and again, so we need to add discount factor to avoid infinity and infinitesimal value function,By introducing discount factor λ, an infinite dimensional problem is transformed into a finite dimensional problem, $0 < \lambda \leq 1$.

$$\begin{aligned}\mathrm{J}(t) = \sum_{l=t}^{\infty} \lambda^{l-t}\mathrm{Y}(x(l), u(l)) = \mathrm{Y}(t) \\ + \lambda \sum_{l=t+1}^{\infty} \lambda^{l-(t+1)}\mathrm{Y}(x(l), u(l))\end{aligned} \tag{4.9}$$

According to the Bellman optimality principle, the optimal value function satisfies:

$$\begin{aligned}\mathrm{J}^{*}(x(t)) = \min_{u(t)}\Big\{x^{T}(t)\mathrm{P}x(t) + 2\int_{0}^{u(t)} \tanh^{-T}(\overline{\mathcal{U}}^{-1}s)\cdot \\ \overline{\mathcal{U}}\mathrm{Q}ds + \lambda\mathrm{J}^{*}(x(t+1))\Big\}.\end{aligned} \tag{4.10}$$

In the ADP algorithm structure, it iterates according to the policy iteration, selecting $T^{\iota}(x)$ as the approximation function and $\o^{\iota}(x)$ as the corresponding control law. The whole iterative process is as follows:

1. Let the initial value function be $T^0(\cdot) = 0$ (which is far from optimal) and compute the control law at "$\iota = 0$"as follows.

$$\o^0(x(t)) = \arg\min_{u(t)}\Big\{x^T(t)Px(t) + 2\int_0^{u(t)} \tanh^{-T}(\overline{\mathcal{U}}^{-1}s) \cdot \overline{\mathcal{U}}Qds + \lambda T^0(x(t+1))\Big\} \tag{4.11}$$

2. Get $T^1(x(t))$:

$$T^1(x(t)) = x^T(t)Px(t) + 2\int_0^{\o^0(x(t))} \tanh^{-T}(\overline{\mathcal{U}}^{-1}s)\overline{\mathcal{U}} \cdot Qds + \lambda T^0(x(t+1)). \tag{4.12}$$

3. And for $\iota = 1, 2, 3, \cdots$

$$\o^{\iota}(x(t)) = \arg\min_{u(t)}\Big\{x^T(t)Px(t) + 2\int_0^{u(t)} \tanh^{-T}(\overline{\mathcal{U}}^{-1}s) \cdot \overline{\mathcal{U}}Qds + \lambda T^{\iota}(x(t+1))\Big\}. \tag{4.13}$$

4. The iterative value function is obtained as follows:

$$T^{\iota+1}(x(t)) = x^T(t)Px(t) + 2\int_0^{\o^{\iota}(x(t))} \tanh^{-T}(\overline{\mathcal{U}}^{-1}s) \cdot \overline{\mathcal{U}}Qds + \lambda T^{\iota}(x(t+1)). \tag{4.14}$$

4.4.2 Structure of Constrained Iterative Dual-Heuristic Dynamic Programming Algorithm

In the dual heuristic dynamic programming, the assumption is that the value function is smooth, modelled on the (4.5), the partial derivatives (4.14) on the right side of $\o^{\iota}(x(t))$, can get [37]:

$$\frac{\partial T^{\iota+1}(x(t))}{\partial u(t)} = \frac{\partial\Big\{x^T(t)Px(t) + 2\int_0^{u(t)} \tanh^{-T}(\overline{\mathcal{U}}^{-1}s)\overline{\mathcal{U}}Qds\Big\}}{\partial u(t)} + \lambda\frac{\partial T^{\iota}(x(t+1))}{\partial u(t)} = 0.$$

And, for $\iota = 0, 1, 2, \cdots$

$$ø^{\iota}(x(t)) = \overline{\mathcal{U}}\tanh\Big(\frac{-\lambda}{2\overline{\mathcal{U}}\mathrm{Q}}\Big[\frac{\partial x(t+1)}{\partial u(t)}\Big]^{T}\frac{\partial \mathrm{T}^{\iota}(x(t+1))}{\partial x(t+1)}\Big) \tag{4.15}$$

Do the same with (4.14) respect to $x(t)$,

$$\frac{\partial \mathrm{T}^{\iota+1}(x(t))}{\partial x(t)} = 2\mathrm{P}x(t) + \lambda\Big[\frac{\partial x(t+1)}{\partial x(t)}\Big]^{T}\frac{\partial \mathrm{T}^{\iota}(x(t+1))}{\partial x(t+1)}. \tag{4.16}$$

As can be seen in (4.15) and (4.16), both have $\frac{\partial \mathrm{T}^{\iota}(x(t+1))}{\partial x(t+1)}$, compared to $\mathrm{T}^{\iota}(x(t))$ in (4.14), DHP algorithm evaluates and updates the first partial derivative of the value function.

The specific algorithm structure is as follows: (set costate function $\mathrm{C}^{\iota}(x(t)) = \partial \mathrm{T}^{\iota}(x(t))/\partial x(t)$).

4.4.3 *Proof of Convergence on I-DHP Algorithm*

The convergence proof of the algorithm shows that with the increase of the number of iterations, the evaluation and update between (4.15) and (4.16) are continuously completed, and the termination condition can finally be satisfied and the optimal solution can be obtained.

The corresponding lemma needs to be given before the formal theorem proving. In order to facilitate writing, abbreviated "$2\int_0^{u(t)} \tanh^{-T}(\overline{\mathcal{U}}^{-1}s)\overline{\mathcal{U}}\mathrm{Q}ds$" to "H(u($t$))".

Lemma 4.1 *Assume that* $ø^{\iota}(t)$ *is the control sequence calculated by (4.13),* $\mathrm{T}^{\iota}(x)$ *is the value function calculated by (4.14).* $!^{\iota}(t)$ *is any admissible control sequence in the domain, and* $\Omega^{\iota}(x)$ *is its corresponding value function equation,*

$$\Omega^{\iota+1}(x(t)) = x^{T}(t)\mathrm{P}x(t) + \mathrm{H}(!^{\iota}(t)) + \lambda\Omega^{\iota}(x(t+1)) \tag{4.17}$$

and it is easy to obtain:

If $\Omega^{0}(\cdot) = \mathrm{T}^{0}(\cdot) = 0$, *then* $0 \le \mathrm{T}^{\iota}(x) \le \Omega^{\iota}(x), \forall\iota.$

Proof The conclusion is obvious. $\mathrm{T}^{\iota}(x)$ is the minimum value that can be obtained on the right side of (4.14), and $ø^{\iota}(t)$ is the corresponding control sequence. And $\Omega^{\iota}(x)$ is any admissible value function, so it must be not less than $\mathrm{T}^{\iota}(x)$. ■

Lemma 4.2 *Given that* $\mathrm{T}^{\iota}(x)$ *by the (4.14), and if the system is controlled, then* $\mathrm{T}^{\iota}(x)$ *has an upper bounded* $\mathfrak{Z}$ *(a constant).*

$$0 \le \mathrm{T}^{\iota}(x) \le \mathfrak{Z}, \forall\iota$$

Algorithm 1: Procedure of the I−DHP algorithm:

INITIAL:

1. Select a smaller positive number ϵ, initial iteration index for $\iota = \mathbf{0}$, $\mathbf{C}^0(\cdot) = \mathbf{0}$.

CALCULATION:

2. Calculate the control law at the **0th** iteration:

$$\begin{aligned}\phi^{\mathbf{0}}(x(t)) &= \arg\min_{u(t)}\Big\{x^T(t)\mathrm{P}x(t) + 2\int_0^{u(t)} \tanh^{-T}(\overline{\mathcal{U}}^{-1}s)\overline{\mathcal{U}}\mathrm{Q}ds + \lambda\mathrm{T}^0(x(t+1))\Big)\\ &= \overline{\mathcal{U}}\tanh\Big(\frac{-\lambda}{2\overline{\mathcal{U}}\mathrm{Q}}\Big[\frac{\partial x(t+1)}{\partial u(t)}\Big]^T\mathbf{C}^{\mathbf{0}}(x(t+1))\Big)\end{aligned}$$

3. Update the costate function for iteration **1**:

$$\mathbf{C}^{\mathbf{1}}(x(t)) = 2\mathrm{P}x(t) + \lambda\Big[\frac{\partial x(t+1)}{\partial x(t)}\Big]^T\mathbf{C}^{\mathbf{0}}(x(t+1))$$

4. Similarly, the control law for the ι iteration:

$$\begin{aligned}\phi^{\iota}(x(t)) &= \arg\min_{u(t)}\Big\{x^T(t)\mathrm{P}x(t) + 2\int_0^{u(t)} \tanh^{-T}(\overline{\mathcal{U}}^{-1}s)\overline{\mathcal{U}}\mathrm{Q}ds + \lambda\mathrm{T}^{\iota}(x(t+1))\Big)\\ &= \overline{\mathcal{U}}\tanh\Big(\frac{-\lambda}{2\overline{\mathcal{U}}\mathrm{Q}}\Big[\frac{\partial x(t+1)}{\partial u(t)}\Big]^T\mathbf{C}^{\iota}(x(t+1))\Big)\end{aligned}$$

5. Obtain the costate function for iteration $\iota + \mathbf{1}$:

$$\mathbf{C}^{\iota+\mathbf{1}}(x(t)) = 2\mathrm{P}x(t) + \lambda\Big[\frac{\partial x(t+1)}{\partial x(t)}\Big]^T\mathbf{C}^{\iota}(x(t+1))$$

COMPARATION:

6. If $\left\|\mathbf{C}^{\iota+\mathbf{1}}(x(t)) - \mathbf{C}^{\iota}(x(t))\right\| \leq \epsilon$, stop and get the approximate optimal control law $\phi^{\iota}(x(t))$;

Else, let $\iota = \iota + \mathbf{1}$, and jump to the **4.**

Proof Set $\upsilon^{\iota}(t)$ to be an admissible and stabilizing control sequence and $\mathcal{V}^{\iota}(x)$ to be:

$$\mathcal{V}^{\iota+1}(x) = x^T(t)\mathrm{P}x(t) + \mathrm{H}(\upsilon^{\iota}(t)) + \lambda\mathcal{V}^{\iota}(x(t+1))$$

Then, it can be obtained: $(\mathcal{V}^0(\cdot) = \mathrm{T}^{\iota}(\cdot) = 0)$

$$\begin{aligned}
\mathcal{V}^{\iota+1}(x) =& x^T(t)\mathrm{P}x(t) + \mathrm{H}(\upsilon^{\iota}(t)) + \lambda\mathcal{V}^{\iota}(x(t+1)) \\
=& x^T(t)\mathrm{P}x(t) + \mathrm{H}(\upsilon^{\iota}(t)) + \lambda\Big[x^T(t+1)\mathrm{P} \\
&\cdot x(t+1) + \mathrm{H}(\upsilon^{\iota-1}(t+1))\Big] + \lambda^2\mathcal{V}^{\iota-1}(x(t+2)) \\
=& \dots \\
=& x^T(t)\mathrm{P}x(t) + \mathrm{H}(\upsilon^{\iota}(t)) + \lambda\Big[x^T(t+1)\mathrm{P} \\
&\cdot x(t+1) + \mathrm{H}(\upsilon^{\iota-1}(t+1))\Big] + \cdots \\
&+ \lambda^{\iota}\Big[x^T(t+\iota)\mathrm{P}x(t+\iota) + \mathrm{H}(\upsilon^0(t+\iota))\Big] \\
&+ \lambda^{\iota+1}\mathcal{V}^0(x(t+\iota+1)).
\end{aligned}$$

$\mathcal{V}^{\iota+l}(x) = \sum_{l=0}^{\iota} \lambda^l \Big[x^T(t+l)\mathrm{P}x(t+l) + \mathrm{H}(\upsilon^{\iota-l}(t+l))\Big] \le \lim_{\iota\to\infty} \Big\{ \sum_{l=0}^{\iota} \lambda^l \Big[x^T(t+l)\mathrm{P}x(t+l) + \mathrm{H}(\upsilon^{\iota-l}(t+l))\Big]\Big\}$.

Due to the admissible control sequence $\upsilon_{\iota}(t)$, it has an upper bound $\mathfrak{Z}$ that

$$\mathcal{V}^{\iota+l}(x) \le \lim_{\iota\to\infty} \Big\{ \sum_{l=0}^{\iota} \lambda^l \Big[x^T(t+l)\mathrm{P}x(t+l) + \mathrm{H}(\upsilon^{\iota-l}(t+l))\Big]\Big\} \le \mathfrak{Z}.$$

Combined with Lemma 4.1, it can be obtained the result. ■

Theorem 4.1 *For the iterative cost function* $\mathrm{T}^{\iota}(x)$ *which follows (4.14) and its corresponding control law* $\phi^{\iota}(t)$ *obtained by (4.13), it can be concluded that with the increase of the number of iterations,* $\mathrm{T}^{\iota}(x)$ *will converge to the optimal value function and* $\phi^{\iota}(t)$ *will converge to the optimal control law, i.e.,* $\mathrm{T}^{\iota}(x) \to \mathrm{J}^*(x)$, $\phi^{\iota}(t) \to u^*(t)$.

Proof From Lemma 4.1, $\Omega^{\iota}(x(t))$ is the cost function corresponding to an any admissible control sequence $!^{\iota}(t)$, with $\Omega^0(\cdot) = 0$.

Firstly, $\iota = 0$,

$$\mathrm{T}^1(x(t)) - \Omega^0(x(t)) = x^T(t)\mathrm{P}x(t) + \mathrm{H}(!^0(t)) \ge 0$$

then, $\mathrm{T}^1(x(t)) \ge \Omega^0(x(t))$, $\iota = 0$.

Secondly, for $\iota - 1$, given $\mathrm{T}^{\iota}(x(t)) \ge \Omega^{\iota-1}(x(t))$, $\forall x(t)$. Then, as ι, it can be able to conclude that

$$\begin{aligned}
\mathrm{T}^{\iota+1}(x(t)) - \Omega^{\iota}(x(t)) =& x^T(t)\mathrm{P}x(t) + \mathrm{H}(!^{\iota+1}(t)) \\
&+ \lambda\big(\mathrm{T}^{\iota}(x(t+1)) - \Omega^{\iota-1}(x(t+1))\big) \\
\ge& \lambda\big(\mathrm{T}^{\iota}(x(t+1)) - \Omega^{\iota-1}(x(t+1))\big)
\end{aligned} \tag{4.18}$$

By the mathematical induction, it can be obtained that $\mathrm{T}^{\iota+1}(x(t)) \geq \Omega^{\iota}(x(t))$, $\forall \iota$. Combined with Lemma 4.1, it is obviously concluded that $\mathrm{T}^{\iota+1}(x(t)) \geq \Omega^{\iota}(x(t)) \geq \mathrm{T}^{\iota}(x(t))$, that is, $\left\{\mathrm{T}^{\iota}(x(t))\right\}$ is a non-decreasing sequence, $\forall \iota$.

From Lemma 4.2, the sequence $\left\{\mathrm{T}^{\iota}(x(t))\right\}$ is bounded to $\mathfrak{Z}$, which is equivalent to that the iterative equation has a limit value, which can be expressed as $\lim_{\iota\to\infty} \mathrm{T}^{\iota}(x(t)) = \mathrm{T}^{\infty}(x(t))$. Therefore, it is bold to assume that $\mathrm{T}^{\infty}(x(t)) = \min\limits_{\o(t)}\left\{x^T(t)\mathrm{P}x(t) + \mathrm{H}(\o(t)) + \lambda\mathrm{T}^{\infty}(x(t))\right\}$. This assumption will be proved below. According to (4.14),

$$\mathrm{T}^{\iota}(x(t)) \leq x^T(t)\mathrm{P}x(t) + \mathrm{H}(\o(t)) + \lambda\mathrm{T}^{\iota-1}(x(t+1)). \tag{4.19}$$

From the non-decreasing property of sequence $\left\{\mathrm{T}^{\iota}(x(t))\right\}$, it can be known that $\mathrm{T}^{\iota}(x(t)) \leq \mathrm{T}^{\infty}(x(t)) \quad \forall \iota$.

Substitute it into (4.19),

$$\mathrm{T}^{\iota}(x(t)) \leq x^T(t)\mathrm{P}x(t) + \mathrm{H}(\o(t)) + \lambda\mathrm{T}^{\infty}(x(t+1)), \quad \forall \iota. \tag{4.20}$$

(4.20) for any ι was established, that when $\iota = \infty$, also meet.

$$\mathrm{T}^{\infty}(x(t)) \leq x^T(t)\mathrm{P}x(t) + \mathrm{H}(\o(t)) + \lambda\mathrm{T}^{\infty}(x(t+1)), \forall \iota. \tag{4.21}$$

Considering that $\o(t)$ is any given control sequence, (4.21) can further obtain:

$$\mathrm{T}^{\infty}(x(t)) \leq \min\left\{x^T(t)\mathrm{P}x(t) + \mathrm{H}(\o(t)) + \lambda\mathrm{T}^{\infty}(x(t+1))\right\}, \forall \iota. \tag{4.22}$$

With (4.14), $\mathrm{T}^{\iota}(x(t)) = \min\limits_{\o(t)}\left\{x^T(t)\mathrm{P}x(t) + \mathrm{H}(\o(t)) + \lambda\mathrm{T}^{\iota-1}(x(t+1))\right\}.\forall \iota$

At this time of its on the left, and as a result of $\left\{\mathrm{T}^{\iota}(x(t))\right\}$ non decreasing, get, $\mathrm{T}^{\infty}(x(t)) \geq \min\limits_{\o(t)}\left\{x^T(t)\mathrm{P}x(t) + \mathrm{H}(\o(t)) + \lambda\mathrm{T}^{\iota-1}(x(t+1))\right\}$. Similarly, let $\iota \to \infty$,

$$\mathrm{T}^{\infty}(x(t)) \geq \min\left\{x^T(t)\mathrm{P}x(t) + \mathrm{H}(\o(t)) + \lambda\mathrm{T}^{\infty}(x(t+1))\right\}, \forall \iota. \tag{4.23}$$

Combining (4.22) and (4.23), it follows that,

$$\mathrm{T}^{\infty}(x(t)) = \min\left\{x^T(t)\mathrm{P}x(t) + \mathrm{H}(\o(t)) + \lambda\mathrm{T}^{\infty}(x(t+1))\right\}, \forall \iota. \tag{4.24}$$

Can be seen from (4.24), the previous assumption proved how. Can be learned from Theorem 4.1, $\mathrm{T}^{\infty}(x(t))$ is a discrete-time time solution of HJB equation. Considering the uniqueness of the solution of the discrete-time-time HJB equation, it means that $T^{\infty}(x(t))$ in (4.24) and $J^{*}(x(t))$ in (4.10) are the same solution. In other words, $\lim_{\iota\to\infty} T^{\iota}(x(t)) = T^{*}(x(t)) = J^{*}(x(t))$. ■

Theorem 4.1 proves that $T^{\infty}(x(t))$ in (4.24) and $J^{*}(x(t))$ in (4.24) in (4.10) are the same solution of the HJB equation corresponding to the same cost function, while the termination criterion "$\left\|T^{\iota+1}(x(t)) - T^{\iota}(x(t))\right\| \leq \epsilon$" indicates that the optimal control law can be solved in finite time, and Theorem 4.2 will explain this context.

Theorem 4.2 *The system (4.1) is controllable and the initial state $x(t)$ of the system can be chosen arbitrarily. Under the finite iteration index ι, the iterative approximate cost function and the optimal cost function $\|T^{*}(x(t)) - T^{\iota}(x(t))\| \leq \epsilon$ are equivalent to the termination criterion $\|T^{\iota+1}(x(t)) - T^{\iota}(x(t))\| \leq \epsilon$.*

Proof In Theorem 4.1, it is mentioned that $\left\{T^{\iota}(x(t))\right\}$ is a non-decreasing sequence, that is

$$J^{*}(x(t)) = T^{*}(x(t)) \geq T^{\iota+1}(x(t)) \geq T^{\iota}(x(t)). \tag{4.25}$$

If $\|T^{*}(x(t)) - T^{\iota}(x(t))\| \leq \epsilon$, it can be concluded that

$$T^{*}(x(t)) - T^{\iota}(x(t)) \leq \epsilon,\ T^{*}(x(t)) \leq T^{\iota}(x(t)) + \epsilon. \tag{4.26}$$

Combined (4.26) with (4.25),

$$T^{\iota}(x(t)) \leq T^{\iota+1}(x(t)) \leq T^{*}(x(t)) \leq T^{\iota}(x(t)) + \epsilon.$$

$$\Rightarrow T^{\iota}(x(t)) \leq T^{\iota+1}(x(t)) \leq T^{\iota}(x(t)) + \epsilon. \tag{4.27}$$

It can get that,

$$\left\|T^{\iota+1}(x(t)) - T^{\iota}(x(t))\right\| \leq \epsilon \tag{4.28}$$

From a different perspective, if (4.28) holds and the $\left\{T^{\iota}(x(t))\right\}$ is nondecreasing,

$$-\epsilon + T^{\iota+1}(x(t)) \leq T^{\iota}(x(t)) \leq T^{*}(x(t)) = J^{*}(x(t)). \tag{4.29}$$

It is obvious that $T^{\iota+1}(x(t)) - T^{*}(x(t) \leq \epsilon$,

$$\left\|T^{\iota+1}(x(t)) - T^{*}(x(t)\right\| \leq \epsilon. \tag{4.30}$$

Based on the analysis of both sides, it can be concluded that $\|T^{*}(x(t)) - T^{\iota}(x(t))\| \leq \epsilon \Leftrightarrow \left\|T^{\iota+1}(x(t)) - T^{\iota}(x(t))\right\| \leq \epsilon$. ■

The two theorems deal with value functions $T(x(t))$, while Algorithm 1 deals with costate function $\mathbf{C}(x(t))$. It will be shown in Theorem 4.3 that this convergence is equivalent.

Theorem 4.3 *(4.14) defines the sequence of value functions. The control law sequence is shown in (4.13) and the update cofunction sequence is shown in (4.16). The optimal value is chosen as the limit of the costate function* $C^*(x(t)) = \lim_{\iota\to\infty} C^{\iota}(x(t))$*, and when the value function approaches the optimal value, the sequence of costate functions converges with the sequence of the control law.*

Proof In Theorems 4.1 and 4.2, it is shown that $T^*(x(t))$ and $T^{\infty}(x(t))$ satisfy the corresponding HJB equation respectively. i.e., $T^{\infty}(x(t)) = T^*(x(t)) = \min_{\o(t)} \Big\{ x^T(t) P x(t) + H(\o(t)) + \lambda T^*(x(t+1)) \Big\}$.

Therefore, it can be concluded that the sequence $\Big\{ T^{\iota}(x(t)) \Big\}$ of value functions converges to the optimal value function of the discrete-time-time HJB equation. i.e.,$T^{\iota} \to T^*$, as $\iota \to \infty$.

Given $C^{\iota}(x(t)) = \partial T^{\iota}(x(t))/\partial x(t)$. It is also possible that the corresponding sequence $\Big\{ C^{\iota}(x(t)) \Big\}$ of costate function converges to $C^{\iota} \to C^*$ as $\iota \to \infty$. Due to the association, costate function is convergent, at the same time, it is concluded that the corresponding sequence converges to the optimal control law $\o^{\iota} \to \o^*$ as $\iota \to \infty$. ■

4.5 Multi-factor Mixed Optimization Experiment Treatment of Tumor Cells

This section explores a novel therapeutic intervention for tumor cell growth inhibition. A discrete-time affine control system has been constructed from the multi-factor tumor cell growth model, and the iterative DHP algorithm has been applied to realize the reduction of drug dosage under the condition of greatly inhibiting the proliferation of tumor cell population.

4.5.1 Discrete Affine Model of Tumor Cell Growth

According to clinical medical statistics and literature [2, 30, 31, 38–41], the values of each parameter in the tumor cell proliferation model affected by multiple factors are shown in Table 4.1.

Using these parameters, try to observe the tumor cell proliferation model given some circumstances.

With reference to [1], the initial "$\mathcal{T}_u(0) = 2 \times 10^7$, $\mathcal{n}_{\hat{k}}(0) = 1 \times 10^3$, $\mathcal{C}_{\mathcal{T}}(0) = 10$, $\mathcal{C}_{\mathcal{L}}(0) = 6 \times 10^8$ " was selected, and the chemotherapy drug at a dose of $\mathcal{V}_{Che}(t) = 3.5$ was injected every 5 days in (4.8) to observe the changes of various cells in the current body.

Table 4.1 Estimated parameter values

Parameter	Description	Estimated value
$\mathfrak{a}$	Tumor cell growth rate	4.31×10^{-1}
$\mathfrak{b}$	The inverse of the carrying capacity of the cell.	1.02×10^{-9}
$\mathfrak{c}_1$	The percentage of circulating lymphocytes that become NK cells	2.08×10^{-7}
$\mathfrak{c}_2$	NK cell mortality	4.12×10^{-2}
$\mathfrak{e}$	$\mathcal{C}_{\mathcal{T}}$ cell mortality	2.04×10^{-1}
$\mathfrak{f}$	Constant source of circulating lymphocytes	7.5×10^{8}
$\mathfrak{g}$	The natural death and differentiation of circulating lymphocyte	1.2×10^{-2}
$\mathfrak{h}_1$	Saturation level of fractional tumor cell kill by $\mathcal{C}_{\mathcal{T}}$ cells	2.34
$\mathfrak{i}$	The killed index of some tumor cells by $\mathcal{C}_{\mathcal{T}}$ cells	2.09
$\mathfrak{j}$	Fractional (non)-ligand-transduced tumor cell killed by NK cells	6.41×10^{-11}
$\mathfrak{l}$	The highest recruitment rate of NK cells in tumor cells	1.25×10^{-2}
$\mathfrak{m}$	Steepness of the NK cell recruitment curve	2.02×10^{7}
$\mathfrak{n}_1$	The rate at which NK cells stimulate $\mathcal{C}_{\mathcal{T}}$ cellproduction after killing tumor cells	1.1×10^{-7}
$\mathfrak{n}_2$	The rate at which stimulate $\mathcal{C}_{\mathcal{T}}$ cell production after killing tumor cells	6.5×10^{-11}
$\mathfrak{p}_1$	Maximum $\mathcal{C}_{\mathcal{T}}$ cell recruitment rate	2.49×10^{-2}
$\mathfrak{p}_2$	NK cell inactivation rate by tumor cells	1×10^{7}
$\mathfrak{q}_1$	Steepness of the $\mathcal{C}_{\mathcal{T}}$ cell recruitment curve	3.66×10^{7}
$\mathfrak{q}_2$	$\mathcal{C}_{\mathcal{T}}$ cell inactivation rate by tumor cells	1.42×10^{-6}
$\mathfrak{r}$	Regulation of $\mathcal{C}_{\mathcal{T}}$ cells by NK cells	3×10^{-10}
$\mathfrak{s}$	Steepness of the $\mathcal{C}_{\mathcal{T}}$ cell recruitment curve	3.66×10^{7}
$\mathfrak{u}$	Regulatory function by NK cells of $\mathcal{C}_{\mathcal{T}}$ cells	3×10^{-10}
$\mathcal{L}_{\mathcal{T}_u}$	Fractional tumor cells are killed by chemotherapy	9×10^{-1}
$\mathcal{L}_{n_{\mathfrak{k}}}$	Fractional NK cells are killed by chemotherapy	6×10^{-1}
$\mathcal{L}_{\mathcal{C}_{\mathcal{T}}}$	Fractional $\mathcal{C}_{\mathcal{T}}$ cells are killed by chemotherapy	6×10^{-1}
$\mathcal{L}_{\mathcal{C}_{\mathcal{L}}}$	Fractional $\mathcal{C}_{\mathcal{L}}$ cells are killed by chemotherapy	6×10^{-1}
γ_α	Rate of chemotherapy drug decay	9×10^{-1}
γ_β	Rate of IL-2 drug decay	1×10^{1}

Figure 4.1 shows an injection method of chemotherapy drugs in the form of pulse. The drug is injected into the body to study the influence of the addition of chemotherapy drugs on the number of various cell populations in the body at different times. As can be seen from the curve of tumor cell change in Fig. 4.1a (the second curve), a dose of 5 chemotherapy drug injected every 5 days for 60 days is sufficient to control the proliferation of tumor cells. The four curves showed different forms of

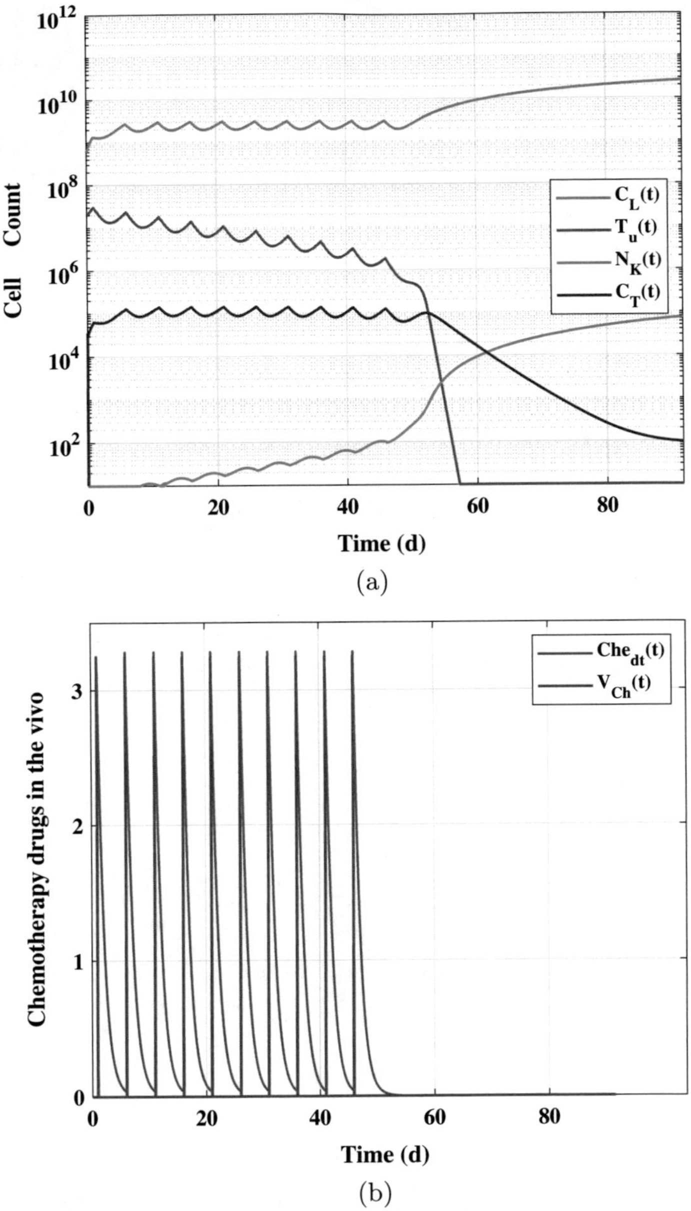

Fig. 4.1 Ten doses of chemotherapy over 60 days has been sufficient to eliminate the tumor. **a** Curves of the population of the four cell species. **b** Distribution of 10 doses of chemotherapy drugs within 60 days and the trend of changes in the concentration of chemotherapy drugs in vivo

oscillatory changes in the early stage, which mainly depended on the pulse injection of chemotherapy drugs, and immunospecific cell $C_{\mathcal{T}}$, which also decreased to stability after tumor cells stabilized in the later stage. Figure 4.1b shows the corresponding mode of administration, with the red is the pulse of administration and the green is the change of the corresponding chemotherapy drug in the body.

4.5.2 Construction of Affine Model

In (4.8), although the discrete model has been obtained, it is too complex and the addition of various coupling forms, which is difficult to be directly combined into the iteration-DHP structure. At this time, the idea of constructing a simple affine model is introduced. It can be easily learned from the above two sub-parts, which can be simplified as the influence of the injected concentrations of the two drugs on tumor cells in the body. Then, the current concentration of tumor cells can be selected as the state variable, and the injected concentrations of the two drugs (chemotherapy drugs and immune agents) can be used as the control variable to form a data set, starting from a large number of random data. The desired affine discrete model is obtained by fitting.

$$x(t+1) = f(x(t)) + \begin{bmatrix} g_1(x(t)) \\ g_2(x(t)) \end{bmatrix}^T u(t) \tag{4.31}$$

$$\begin{aligned} g_1\left(\log_{10}(x)\right) = {} & 0.001771\left(\log_{10}(x)\right)^5 - 0.02931\left(\log_{10}(x)\right)^4 + 0.1793\left(\log_{10}(x)\right)^3 \\ & - 0.5353\left(\log_{10}(x)\right)^2 + 1.741\left(\log_{10}(x)\right) - 1.133 \end{aligned} \tag{4.32}$$

$$\begin{aligned} g_2\left(\log_{10}(x)\right) = {} & 0.007579\left(\log_{10}(x)\right)^4 - 0.1087\left(\log_{10}(x)\right)^3 \\ & + 0.4838\left(\log_{10}(x)\right)^2 + 0.1783\left(\log_{10}(x)\right)^2 - 0.2304 \end{aligned} \tag{4.33}$$

4.5.3 Optimization of Mixed Treatment Regimen

Following the affine model mentioned above, it is necessary to specify the cost function required in iteration-DHP before optimizing the treatment:

Table 4.2 Default parameters

λ	P	m_1	$\overline{\mathcal{U}}_1$	$\overline{\mathcal{U}}_1$	m_2	$\overline{\mathcal{U}}_2$	Q_2
0.95	10^{-7}	0.55	1.28	6×10^48	0.62	1.12	2×10^5

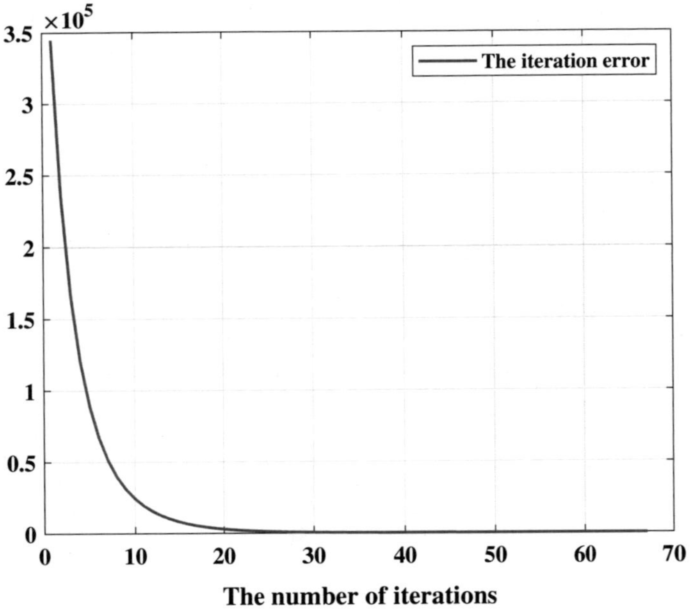

Fig. 4.2 The iteration error change curve, after the end of the 67th iteration, satisfies the termination condition

$$J(x(t)) = \sum_{\iota=0}^{\infty} \lambda^{\iota} \Big\{ x^T(t) P x(t) + m_1 \int_0^{u_1(t)} \tanh^{-T}(\overline{\mathcal{U}}_1^{-1} s) \cdot \overline{\mathcal{U}}_1 Q_1 ds + m_2 \int_0^{u_2(t)} \tanh^{-T}(\overline{\mathcal{U}}_2^{-1} s) \overline{\mathcal{U}}_2 Q_2 ds \Big\}. \tag{4.34}$$

According to clinical experience, the default parameters are shown in Table 4.2. The iteration error ϵ is set to 10^{-6}, and the iteration error variation curve is shown in Fig. 4.2. The error decreases extremely fast in the first twenty iterations of the calculation, and the convergence rate gradually decreases after 20 iterations. At $\iota = 67$, the termination condition has been satisfied.

Analysis of tumor cells after meet the termination criterion, according to the optimized regimen of population change curve as shown in Fig. 4.3, visible at an extremely rapid rate by the growth of stem. The usage and dosage of two drugs

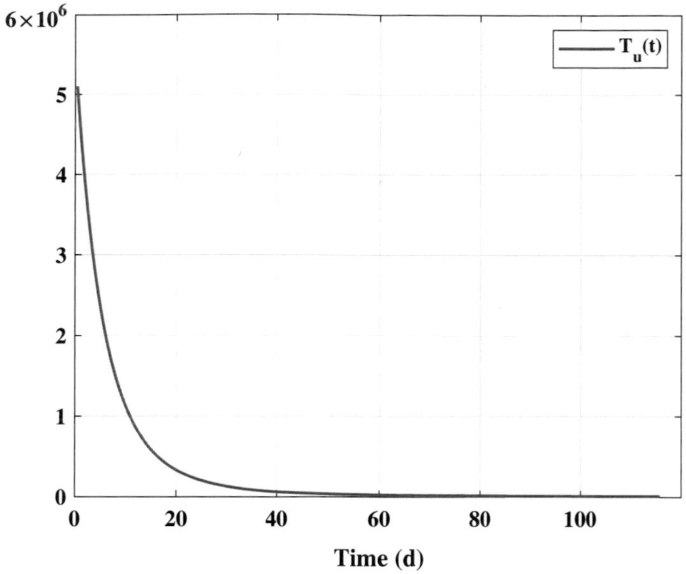

Fig. 4.3 Tumor cell population changes in optimized treatment

are shown in Fig. 4.4. Figure 4.4a represents the curve of injected concentration of chemotherapy drugs, and Fig. 4.4b represents the curve of injected concentration of immune drugs.

4.6 Conclusion

In this chapter, a tumor immune differential game system has been established to solve the problem of optimal clinical tumor treatment oriented to evolutionary dynamics. Firstly, a mathematical model of the game system between tumor cells and immune cells treated by immune agents and chemotherapy drugs has been given. Secondly, the bounded optimal control problem has been solved by the HJB equation with infinite horizon performance index which is subjected to practical constraints. Finally, the optimal iterative approximate control strategy has been obtained by the iterative dual heuristic dynamic programming algorithm, and the effectiveness of the proposed algorithm has been proved.

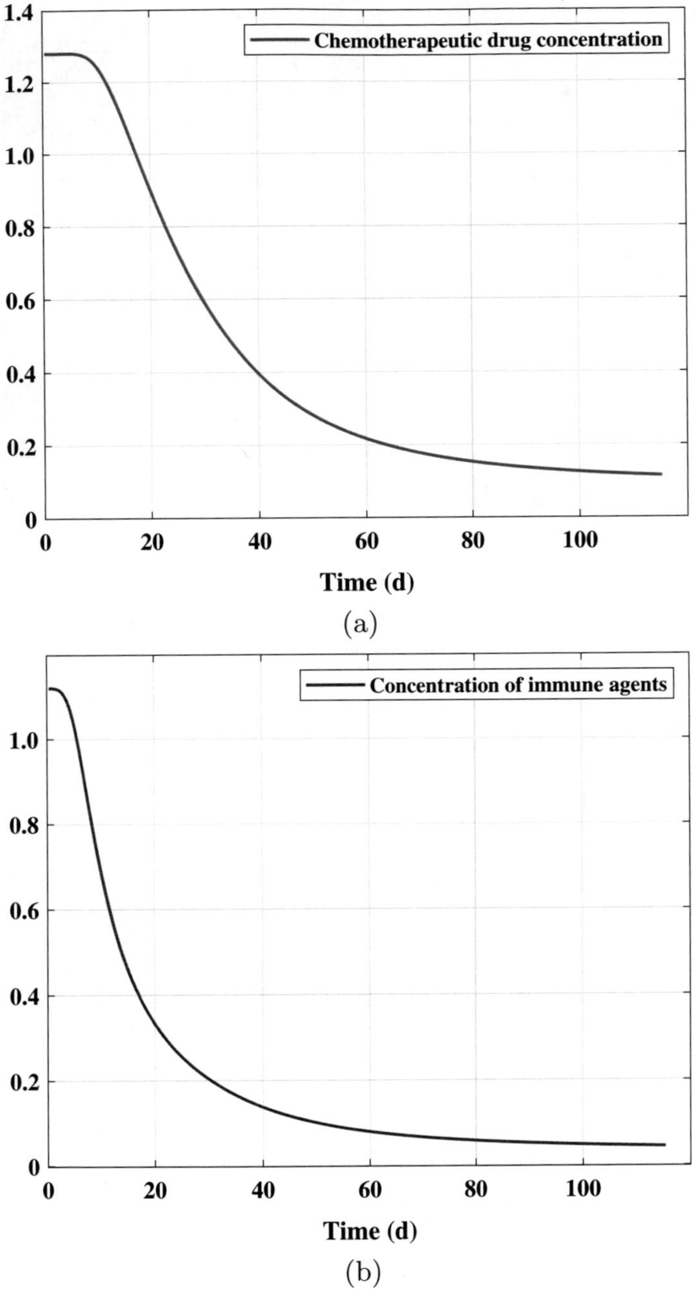

Fig. 4.4 Optimization of treatment of different drugs usage and dosage. **a** Injection concentration of chemotherapeutic drugs, **b** Injection concentration of immune agents

References

1. Gao H, Li W, Pan M, Han Z, Poor HV (2021) Modeling COVID-19 with mean field evolutionary dynamics: social distancing and seasonality. J Commun Netw 23(5):314–325
2. Diefenbach A, Jensen ER, Jamieson AM, Raulet D (2001) Rae1 and H60 ligands of the NKG2D receptor stimulate tumor immunity. Nature 413(6852):165–171
3. Cassetta L, Pollard JW (2018) Targeting macrophages: therapeutic approaches in cancer. Nat Rev Drug Discov 17(12):887–904
4. Yang Y, Modares H, Vamvoudakis KG, He W, Xu CZ, Wunsch DC (2022) Hamiltonian-driven adaptive dynamic programming with approximation errors. IEEE Trans Cybernet 52(12):13762–13773
5. Rizvi SAA, Lin Z (2022) Adaptive dynamic programming for model-free global stabilization of control constrained continuous-time systems. IEEE Trans Cybernet 52(2):1048–1060
6. Luo S, Lewis FL, Song Y, Ouakad HM (2022) Optimal synchronization of unidirectionally coupled FO chaotic electromechanical devices with the hierarchical neural network. IEEE Trans Neural Netw Learn Syst 33(3):1192–1202
7. Yang Y, Vamvoudakis KG, Modares H, Yin Y, Wunsch DC (2021) Hamiltonian-driven hybrid adaptive dynamic programming. IEEE Trans Syst Man Cybernet: Syst 51(10):6423–6434
8. Moghadam R, Natarajan P, Jagannathan S (2022) Online optimal adaptive control of partially uncertain nonlinear discrete-time systems using multilayer neural networks. IEEE Trans Neural Netw Learn Syst 33(9):4840–4850
9. Luo B, Liu D, Huang T, Wang D (2016) Model-free optimal tracking control via critic-only Q-learning. IEEE Trans Neural Netw Learn Syst 27(10):2134–2144
10. Niu H, Bhowmick C, Jagannathan S (2020) Attack detection and approximation in nonlinear networked control systems using neural networks. IEEE Trans Neural Netw Learn Syst 31(1):235–245
11. Yang X, He H, Zhong X (2021) Approximate dynamic programming for nonlinear-constrained optimizations. IEEE Trans Cybernet 51(5):2419–2432
12. Zhang D, Ye Z, Feng G, Li H (2022) Intelligent event-based fuzzy dynamic positioning control of nonlinear unmanned marine vehicles under DoS attack. IEEE Trans Cybernet 52(12):13486–13499
13. Huang M, Jiang ZP, Ozbay K (2022) Learning-based adaptive optimal control for connected vehicles in mixed traffic: robustness to driver reaction time. IEEE Trans Cybernet 52(6):5267–5277
14. Sun J, Zhang H, Wang Y, Sun S (2022) Fault-tolerant control for stochastic switched IT2 fuzzy uncertain time-delayed nonlinear systems. IEEE Trans Cybernet 52(2):1335–1346
15. Liu P, Sun J, Zhang H, Xu S, Liu Y (2023) Combination therapy-based adaptive control for organism using medicine dosage regulation mechanism. IEEE Trans Cybernet. https://doi.org/10.1109/TCYB.2022.3196003
16. Liu D, Baldi S, Yu W, Chen G (2022) On distributed implementation of switch-based adaptive dynamic programming. IEEE Trans Cybernet 52(7):7218–7224
17. Labao AB, Martija MAM, Naval PC (2021) A3C-GS: adaptive moment gradient sharing with locks for asynchronous actor-critic agents. IEEE Trans Neural Netw Learn Syst 32(3):1162–1176
18. Moghadam R, Jagannathan S (2023) Optimal adaptive control of uncertain nonlinear continuous-time systems with input and state delays. IEEE Trans Neural Netw Learn Syst. https://doi.org/10.1109/TNNLS.2021.3112566
19. Al-Dabooni S, Wunsch DC (2020) An improved n-step value gradient learning adaptive dynamic programming algorithm for online learning. IEEE Trans Neural Netw Learn Syst 31(4):1155–1169
20. Wang D, Liu D, Mu C, Zhang Y (2018) Neural network learning and robust stabilization of nonlinear systems with dynamic uncertainties. IEEE Trans Neural Netw Learn Syst 29(4):1342–1351

21. Wang D, He H, Liu D (2017) Adaptive critic nonlinear robust control: a survey. IEEE Trans Cybernet 47(10):3429–3451
22. Britton NF (2003) Essential mathematical biology. Springer, Berlin
23. Yazdani D, Cheng R, He C, Branke J (2022) Adaptive control of subpopulations in evolutionary dynamic optimization. IEEE Trans Cybernet 52(7):6476–6489
24. Mu C, Peng J, Sun C (2023) Hierarchical multiagent formation control scheme via actor-critic learning. IEEE Trans Neural Netw Learn Syst. https://doi.org/10.1109/TNNLS.2022.3153028
25. Wang D, Hu L, Zhao M, Qiao J (2023) Adaptive critic for event-triggered unknown nonlinear optimal tracking design with wastewater treatment applications. IEEE Trans Neural Netw Learn Syst. https://doi.org/10.1109/TNNLS.2021.3135405
26. Werbos PJ (1992) Approximate dynamic programming for real-time control and neural modeling. Handbook of intelligent control. Van Nostrand, New York, NY, USA
27. Prokhorov DV, Santiago RA, Wunsch DC II (1995) Adaptive critic designs: a case study for neurocontrol. Neural Netw 8(9):1367–1372
28. de Pillis LG, Gu W, Radunskaya AE (2006) Mixed immunotherapy and chemotherapy of tumors: modeling, applications and biological interpretations. J Theor Biol 238(4):841–862
29. de Pillis LG, Radunskaya AE (2003) Immune response to tumor invasion. Comput Fluid Solid Mech 2:1661–1668
30. Kirschner D, Panetta JC (1998) Modeling immunotherapy of the tumor-immune interaction. J Math Biol 37(3):235–252
31. Kuznetsov V, Makalkin I, Taylor M, Perelson A (1994) Nonlinear dynamics of immunogenic tumors: parameter estimation and global bifurcation analysis. Bull Math Biol 56(2):295–321
32. Huang AYC, Golumbek P, Ahmadzadeh M, Jaffee E, Pardoll D, Levitsky H (1994) Role of bone marrow-derived cells in presenting MHC class I-restricted tumor antigens. Science 264(5161):961–965
33. Gilbertson SM, Shah PD, Rowley DA (1986) NK cells suppress the generation of Lyt-2+ cytolytic T cells by suppressing or eliminating dendritic cells. J Immunol 136(10):3567–3571
34. Gett A, Sallusto F, Lanzavecchia A, Geginat J (2003) T cell fitness determined by signal strength. Nat Immunol 4(4):355–360
35. Rosenberg S, Lotze M (1986) Cancer immunotherapy using interleukin-2 and interleukin-2-activated lymphoytes. Ann Rev Immunol 4:681–709
36. Gardner SN (2000) A mechanistic, predictive model of dose-response curves for cell cycle phase-specific and nonspecific drugs. Can Res 60(5):1417–1425
37. Dierks T, Thumati BT, Jagannathan S (2009) Optimal control of unknown affine nonlinear discrete-time systems using offline-trained neural networks with proof of convergence. Neural Netw 22(5–6):851–860
38. Dudley ME, Wunderlich JR... Rosenberg SA (2002) Cancer regression and autoimmunity in patients after clonal repopulation with antitumor lymphocytes. Science 298(5594):850–854
39. Kuznetsov V, Makalkin I (1992) Bifurcation-analysis of mathematical-model of interactions between cytotoxic lymphocytes and tumor-cells-effect of immunological amplification of tumor-growth and its connection with other phenomena of oncoimmunology. Biofizika 37(6):1063–1070
40. Yates A, Callard R (2002) Cell death and the maintenance of immunological memory. Discret Contin Dyn Syst Ser B 1(1):43–59
41. Lanzavecchia A, Sallusto F (2000) Dynamics of T-lymphocyte responses: intermediates, effectors, & memory cells. Science 290(5489):92–97

Chapter 5
N-Level Hierarchy-Based Optimal Control to Develop Therapeutic Strategies for Ecological Evolutionary Dynamics Systems

5.1 Introduction

The death toll from tumor diseases is on the rise, and the nonlinear dynamics and control of tumor growth have attracted widespread attention [1]. The number of tumor cells is gradually increasing. The most obvious feature is abnormal anti-growth signals. There is a strict control mechanism for normal cells. However, in the continuous process, the static and death signals are turned off to generate cell division signals, which leads to the crazy growth of tumor cells [2, 3]. Tumor cells promote the growth of blood vessels, which are necessary to provide nutrients. This is why the flow of blood in tumor tissues is related to the benign or malignant tumor. Cancer cells are also polarized. They have evolved their camouflage ability in the ongoing battle with immune cells, causing the immune system to mistake them for normal cells, which makes it difficult for chemotherapeutic drugs to distinguish the volume of biological targets [4, 5]. When the differentiation process of normal cells is not controlled, they will evolve into tumor cells. This is the nature of tumor cells, tumor cells continue to proliferate, deprive their limited body energy supply, and ultimately destroy the body's function and die [6]. Therefore, in order to inhibit the growth of tumor cells, it is urgent to find a treatment that will minimize the damage to oneself.

In the fight against cancer, before the advent of chemotherapy and radiotherapy, there have been no effective measures for the small differences between cancer cells and normal cells [7, 8]. When the side effects of radiotherapy and chemotherapy increased and the targeted therapy was highly targeted and inflexible, scientific research projects began to turn to humans themselves [9]. The complex and unique communities of cell life are called microenvironments by scientists. The microenvironment has many characteristics that affect cell growth, behavior, and how to communicate with other cells nearby [10]. In the oncology world, researchers are committed to understanding the tumor microenvironment and trying to find feasible treatment opportunities. Under normal circumstances, the immune system can recognize and eliminate tumor cells in the tumor microenvironment. However, in order to survive and grow, tumor cells can adopt different strategies to suppress the body's

J. Sun et al., *Adaptive Dynamic Programming*,
https://doi.org/10.1007/978-981-99-5929-7_5

immune system and fail to kill tumor cells normally, thereby surviving the various stages of the anti-tumor immune response. The above-mentioned characteristics of tumor cells are called immune escape. Tumor cells escape the immune system, not because the immune system cannot recognize them, nor because it is not activated, but cancer cells have evolved a way to prevent T cell activation through specific binding [11–13]. Therefore, the medical community has been trying to find many special methods to treat cancer cells to block the activation of T cells and release the immune system.

Chemotherapy not only kills rapidly differentiated tumor cells, but also involves conventional cells. Its side effects are the most obvious, but they can be alleviated by immunotherapy. The closure of immune checkpoints and the success of adoptive cell therapy have made immunotherapy a mature means of treating cancer [14, 15]. Compared with traditional therapies such as surgery, radiotherapy, and chemotherapy, immunotherapy has fewer side effects and better effects, but immunotherapy is difficult to overcome its transient nature. With the rapid increase in tumor patients, immunotherapy is rapidly emerging for the treatment of specific types of cancer, especially tumors with poor immunogenicity [2]. The original intention of immunotherapy is to fight cancer cells through the lethality of immune cells themselves. As a typical immune deficiency syndrome, AIDS is caused by the failure of the immune response and is often attributed to the weakening of the immune level. However, once the activated immune system cannot be stopped, cytokines are produced, which is considered to be an overreaction of the immune system like COVID-19 [16, 17]. Therefore, the combined treatment of chemotherapy and immunotherapy is more reasonable. Immunotherapy refers to a treatment method that artificially enhances or suppresses the body's immune function to achieve the purpose of curing diseases by referring to the body's low or hyperimmune state. There are many immunotherapy methods, which are suitable for the treatment of many diseases. Tumor immunotherapy aims to activate the human immune system, relying on its own immune function to kill cancer cells and tumor tissues [18]. Unlike previous surgery, chemotherapy, radiotherapy, and targeted therapy, the target of immunotherapy is not tumor cells and tissues, but the body's own immune system [19]. Different types of tumor cells interact with different types of immune cells, and these immune cells have the function of helping or attacking tumors [20].

The mechanism of immune regulation varies from person to person, but in the case of special calls, the optimal regulation based on immunotherapy will play a role in reducing tumor cells regardless of specific circumstances. Enhancing tumor antigen presentation can effectively stimulate dendritic cells and improve immunotherapeutic efficacy [21, 22]. The known "predation-prey" between immune cells and tumor cells will cause periodic growth and reduction of cells. This growth and reduction can continue indefinitely or reach a balanced saddle point determined by system parameters [23]. And all of the above is composed of a complex non-linear structure, and it is difficult to achieve the global optimum with conventional optimization methods. Especially for the treatment of the human body, how to rationally use drugs to achieve the minimum harm to the human body is particularly important. So this article proposes a novel evolutionary calculation method, N-Level Hierarchy Optimization (NLHO) algorithm. It is bionic from the hierarchical system of biological populations in the natural world. The hierarchical system refers to the hierarchical phenomenon in which the status of each animal in the animal group has a certain order. The basis of the formation of the hierarchy is the dominance behavior, that is, the "domination-submission" relationship [24]. When the formed hierarchical system stabilizes, lower-ranking people generally show compromise and obedience, but sometimes they also re-struggle to change the hierarchical order, and so on. A stable population will develop for a long time. This is an explanation for the rationality of the hierarchy preserved in evolutionary selection [25]. So for the entire species population, this is conducive to the preservation and continuation of the species. A variety of biological interactions constitute a complex nonlinear growth process of tumor cells, and the main influencing factors of tumor cell populations are the focus of research. Hunting cells refer to immune cells that participate in the removal of foreign objects and strengthen the immune response [26, 27].

In the NLHO algorithm, an N levels optimization structure is designed, which includes the leader level, guider level, executant level and follower level. In the entire population, the individual with the best search position is selected as the leader, who has the grasp of the entire search direction of the team it leads. The second level is the guider level, which executes the tasks issued by the leader and follows the direction of the leader to find the best. Of course, in the whole process, the guider will also refer to the task allocation of the global optimal leader to guide the executants to find the best, so as to prevent the leader of the team from falling into a local optimum. The third level is the executant level, which follows the guider to complete the task, in order to achieve a wider area of coverage search. At the same time, it will also refer to the tasks assigned by the leaders of the ethnic group to make the task goals clearer and speed up the convergence. The last level is the follower level. At this level, followers can be divided into any level to solve different optimization problems. Of course, in the later stage of searching, there may be excessive overlap between population individuals [28].

5.2 Ecological Evolutionary Dynamics Systems Model

This part mainly introduces the mathematical growth model of tumor cells, which takes into account the influence of external factors such as chemotherapy drugs and immunotherapy on tumor cells, as well as the interaction between the two cells. In the following model, $T(\mathrm{t})$ represents the number of tumor cells, $I(\mathrm{t})$ represents the number of immune cells, $Con_{che}(t)$ and $Con_{im}(t)$ represent the blood concentration of chemotherapy drugs and immunotherapy drugs, respectively.

Taking into account the interaction between immune cells and tumor cells, the direct killing of chemotherapeutic drugs and the growth model of tumor cells can be written as

$$T(t+1) = T(t) + \vartheta_1 \times T(t) \times \big(1 - \vartheta_2 \times T(t)\big) \\ - \gamma \times T(t) \times I(t) - \varepsilon \times T(t) \times Con_{che}(t) \tag{5.1}$$

where, ϑ_1 stands for inherent growth rate unrelated to immune cells and chemotherapy drugs, ϑ_2 stands for the maximum interaction ability between immune cells and tumor cells, ignoring chemotherapy drugs, γ stands for the growth rate when tumor cells are inactivated and attacked by immune cells, ε stands for the stress response coefficient of tumor cells to chemotherapeutics.

Considering the natural growth law of immune cells, we assume that a fixed number of immune cells are produced in a unit time, and these cells have an inevitable life cycle. Tumor cells in the body can stimulate the growth of immune cells, which is a positive non-linear change. In immunotherapy, the addition of immune drugs can produce an immune response, leading to non-linear growth of immune cells. At the same time, in the struggle between immune cells and tumor cells, the immune cells themselves will also cause losses. In chemotherapy, chemotherapy drugs can also cause damage to immune cells.

$$I(t+1) = I(t) + \vartheta_3 - \lambda \times I(t) \\ + \frac{\alpha_1 \times T^2(t) \times I(t)}{\beta_1 + T^2(t)} + \frac{\alpha_2 \times T(t) \times \mathrm{Con}_{\mathrm{im}}(t)}{\beta_2 + \mathrm{Con}_{\mathrm{im}}(t)} \\ - \xi_1 \times T(t) \times I(t) - \xi_2 \times \mathrm{Con}_{\mathrm{che}}(t) \times I(t) \tag{5.2}$$

where, ϑ_3 stands for rate of continuous inflow, λ stands for natural decay rate without any external effects, α_1 stands for maximum recruitment rate caused by tumor cells, α_2 stands for the largest proportion of tumor cells caused by immunotherapeutic drugs, β_1 stands for steepness factor caused by tumor cells, β_1 stands for steepness coefficient caused by immunotherapeutic drugs, ξ_1 stands for stress response coefficient to chemotherapy drugs, ξ_2 stands for response rate of tumor cells to immune cells.

At a certain point in time after the injection of chemotherapy drugs, the concentration of the drugs in the body will decrease exponentially. We are adding immune drugs at the same time. We can get the attenuation model of chemotherapy drugs and immune drugs in vivo.

$$Con_{che}(t+1) = \chi_{che}(t) - e^{-\theta_1} Con_{che}(t) \tag{5.3}$$

$$Con_{im}(t+1) = \chi_{im}(t) - e^{-\theta_2} Con_{im}(t) \tag{5.4}$$

where, $\chi_{che}(t)$ and $\chi_{im}(t)$ represent the concentration of chemotherapeutic drugs and immune drugs, respectively. θ_1 and θ_2 are the attenuation rates of chemotherapy drugs and immune drugs.

When we qualitatively analyze how to minimize the number of tumor cells remaining in the bloodstream under the premise of using as few drugs as possible, including chemotherapy drugs and immune drugs, this process can be described by quantitative mathematical expressions. From formulas (5.1)–(5.4), we can get:

$$F_{\min} = \sum_{t=t_0}^{t} \delta^t \left\{ \omega T^2(t) + \int_0^{\chi_{\text{che}}(t)} \tan^{-1}(\bar{U}_1^{-1} s)\bar{U}_1 R_1 ds + \int_0^{\chi_{\text{im}}(t)} \tan^{-1}(\bar{U}_2^{-1} s)\bar{U}_2 R_2 ds \right\} \tag{5.5}$$

where, $\bar{U}_1$ and $\bar{U}_2$ respectively represent the maximum allowable dose of chemotherapy drugs and the dose of a single injection of immunizing agent, δ is the discount factor, ω is a constant coefficient.

5.3 N-Level Hierarchy Optimization Algorithm

5.3.1 *Leader Level of the Hierarchy*

First of all, as individuals with high fitness values, leaders have strong self-learning capabilities. Therefore, the iterative formula of design leaders is as follows:

$$x_{l,j}^{t+1} = x_{l,j}^{t}\left(1 + randn(\mu_l, \sigma_l)\right) \tag{5.6}$$

where, i denotes the ith leader in the population, and j is the dimension. t is the number of iterations. Randn is a Gaussian distribution, where the mean $\mu_l = 0$ and the standard deviation σ_l is shown below:

$$\sigma_l = \begin{cases} 1 & , f_l^t \le f_i^t \\ exp(f_l^t - f_i^t) & , f_l^t > f_i^t \end{cases}, i \in [1, 2, \cdots N], i \neq l \tag{5.7}$$

where, f_l^t is the fitness value of the lth leader at the tth iteration, and f_i^t is the fitness value of any individual in the population that is different from the lth leader.

5.3.2 *Guider Level of the Hierarchy*

Secondly, as the individuals who guide the general direction of the evolution of the entire population for the leader, the guider must not only learn from the best overall, but also obey the leader's command.

$$x_{g,j}^{t+1} = x_{g,j}^{t} + randn(\mu_g, \sigma_g^2) \times (x_{l,j}^{t} - x_{g,j}^{t}) + s_1 \times (x_{best,j}^{t} - x_{g,j}^{t}) \tag{5.8}$$

where, g denotes the gth guider in the population, best is the best individual in the current iteration, s_1 is the acceptance factor of guider, $\mu_g = 0.5$.

$$\sigma_g = exp(f_l^t - f_g^t) \tag{5.9}$$

$$s_1 = exp\left(\frac{f_{best}^t - f_g^t}{|f_g^t| + \varepsilon}\right) \tag{5.10}$$

where, ε is an infinitesimal value to prevent a guider from having a fitness value of 0.

5.3.3 Executant Level of the Hierarchy

The executants seek the best as the main body of the entire population. On the one hand, follow the guider's arrangements, and on the other hand, follow the leader's direction.

$$x_{e,j}^{t+1} = x_{e,j}^{t} + randn(\mu_e, \sigma_e^2) \times (x_{g,j}^{t} - x_{e,j}^{t}) + s_2 \times (x_{l,j}^{t} - x_{e,j}^{t}) \tag{5.11}$$

where, e is each executant in the population, s_2 is the acceptance factor of the executor, $\mu_e = 0.8$.

$$\sigma_e = exp\left(\frac{f_g^t - f_e^t}{|f_e^t| + \varepsilon}\right) \tag{5.12}$$

$$s_2 = exp(f_l^t - f_e^t) \tag{5.13}$$

5.3.4 Follower Level of the Hierarchy

Finally, there are followers, who themselves will be divided into multiple levels. Learn from each other at different levels, and notify the follow-up executant to check for deficiencies.

$$x_{f_n,j}^{t+1} = x_{f_n,j}^{t} + randn(\mu_{f_n}, \sigma_{f_n}^2) \times (x_{f_{n-1},j}^{t} - x_{f_n,j}^{t}) + c_n \times rand \times (x_{e,j}^{t} - x_{f_n,j}^{t}) \tag{5.14}$$

where, f_n is the n-level follower, c_n is the absorption factor of the n-level follower, $\mu_{f_n} = 0.8 - 0.6 \times (t/t_{max})$,$x^t_{f_0,j} = x^t_{e,j}$,$f^t_{f_0} = f^t_e$, n is a natural number greater than 0.

$$\sigma_{f_n} = exp(f^t_{f_{n-1}} - f^t_{f_n}) \tag{5.15}$$

$$c_n = exp(f^t_{f_{n-1}} - f^t_{f_n}) \tag{5.16}$$

5.4 Simulation and Analysis for NLHO

In this experiment, the population size is set to 100, and the maximum number of iterations is set to be 100. Each algorithm is run independently for 50 times, and the spatial dimension is selected according to different test functions. The distribution rates of each level system are LPercent = 10%, GPercent = 20%, EPercent = 40%, and FPercent = 30%. The value of the updated algebra G = 10.

For independent tests of 20 test functions, we separately count their mean, minimum and standard deviation to evaluate the performance of NLHO in various aspects by setting the difficulty in different aspects. At the same time, we select some typical algorithms for comparison, such as Taboo Search (TS), Chicken Swarm Optimization (CSO), Genetic Algorithm (GA), Ant Colony Optimization (ACO), and Simulated Annealing (SA), so as to compare the performance of the NLHO algorithm horizontally. The test results are shown in Table 5.1.

For the independent tests of the benchmark functions, we calculated 5 parameter indexes respectively, which were their best, worst, median, average and std. deviation. At the same time, in order to verify whether the results are statistically significant, we use the Wilcoxon rank-sum test between NLHO and the other algorithms. "+",

Table 5.1 Experimental simulation results

Function	Optimization	TS	CSO	GA	ACO	SA	NLHO
Ackley	Mean	2.685462(–)	**8.88E-16**(≈)	17.98669(–)	8.157645(–)	0.381913(–)	**8.88E-16**
	Min	1.599236(–)	**8.88E-16**(≈)	2.532499(–)	0.818328(–)	0.024938(–)	**8.88E-16**
	Std. Deviation	3.327767(–)	**0**(≈)	5.021448(–)	2.860999(–)	0.248066(–)	**0**
Cross-in-Tray	Mean	−2.04281(–)	−2.06247(–)	−1.52287(–)	−2.05567(–)	−2.05048(–)	**−2.06261**
	Min	−2.05463(–)	**−2.06261**(≈)	−2.06257(–)	−2.06261(–)	−2.06257(–)	**−2.06261**
	Std. Deviation	0.050988(–)	0.000129(–)	0.314647(–)	0.007304(–)	0.008585(–)	**1.17E-10**
Drop-Wave	Mean	−0.90552(–)	**−1**(≈)	−0.16478(–)	−0.9568(–)	−0.98441(–)	**−1**
	Min	−0.93563(–)	**−1** (≈)	−0.90648(–)	**−1**(≈)	−0.99988(–)	**−1**
	Std. Deviation	0.105019(–)	1.63E-07(–)	0.194473(–)	0.025337(–)	0.016453(–)	**0**

"–", and "≈" mean that the proposed NLHO is significantly better, significantly worse, and no significantly statistically different in the comparison, respectively.

It can be seen from the experimental results that, compared with the other 5 algorithms, under 20 test functions 60 indicators, NLHO wins 58, 29, 59, 53 and 56 indicators, and they belong to different types of test functions, which reflects the better robustness of NLHO. Moreover, it can be seen from the Wilcoxon rank-sum test that NLHO is not only different from other population algorithms, but also has obvious advantages. The experimental results for each function are discussed in more detail below.

The Ackley function f_1 is widely used to test optimization algorithms. It is a continuous experimental function obtained by superimposing an exponential function with a moderately amplified cosine. It is characterized by an almost flat outer area, which is modulated by a cosine wave to form holes or peaks, making the curved surface undulating, but there is a large hole in its center. For the Ackley function, both NLHO and CSO have found the global minimum, and the average value is also equal to the global minimum, so that the standard deviation is also 0. The optimization process diagram of the NLHO algorithm is shown in this article, including the initial iteration diagram, the final result diagram and the intermediate process convergence curve, as shown in Fig. 5.1. The two algorithms are comparable, SA performance is average, while TS, GA and ACO perform poorly.

The Cross-in-Tray function f_2 has multiple global minimums. On this function, the mean, minimum and standard deviation of NLHO have reached the best, and the experimental results are shown in Fig. 5.2. At the same time, ACO also found the global optimum, but the mean and standard deviation are slightly inferior to NLHO. TS, CSO and SA performed well, while GA performed average.

The Drop-Wave function f_3 is multi-modal and very complicated. In each smaller input domain, its features have multiple ring-shaped peaks and valleys, and the depth of the valleys gradually decreases as the center of the circle shrinks. For the optimization process, it is very easy to fall into the local optimum. For this kind of complex optimization function, NLHO has shown strong optimization ability, and the global minimum, mean and standard deviation have all reached the best. The experimental results are shown in Fig. 5.3. CSO, ACO and SA performed well, and TS and GA performed poorly.

5.5 Develop Therapeutic Strategies for Ecological Evolutionary Dynamics Systems Using NLHO

In this section, we apply the NLHO algorithm to the EEDS model as an experimental verification. According to clinical treatment needs, chemotherapeutic drugs and immune drugs are used as input, and the cost of treatment loss is used as the objective function. Through the iteration of the NLHO algorithm, the optimal therapeutic strategies for patients with a certain basic condition are worked out. For some of the

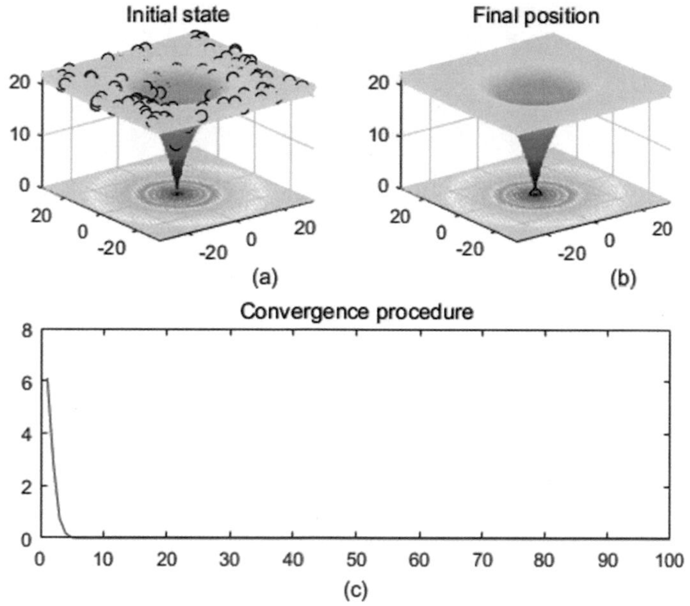

Fig. 5.1 Ackley

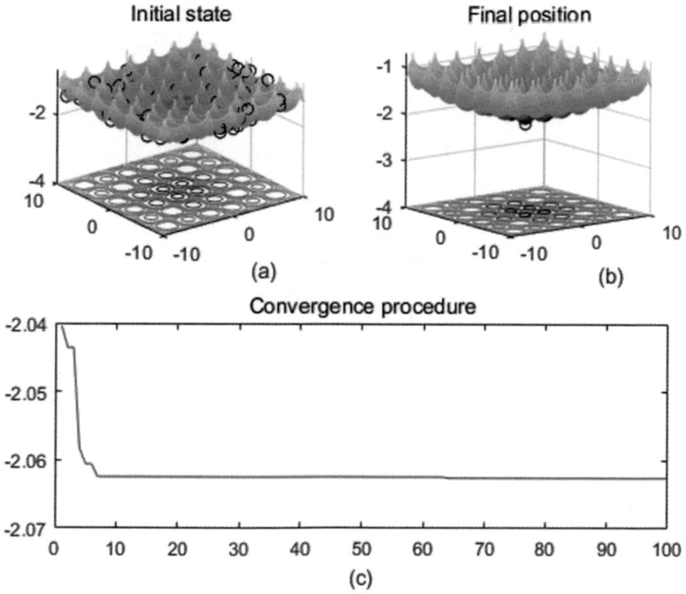

Fig. 5.2 Cross-in-tray

remainder of this sample we will use dummy text to fill out paragraphs rather than use live text that may violate a copyright.

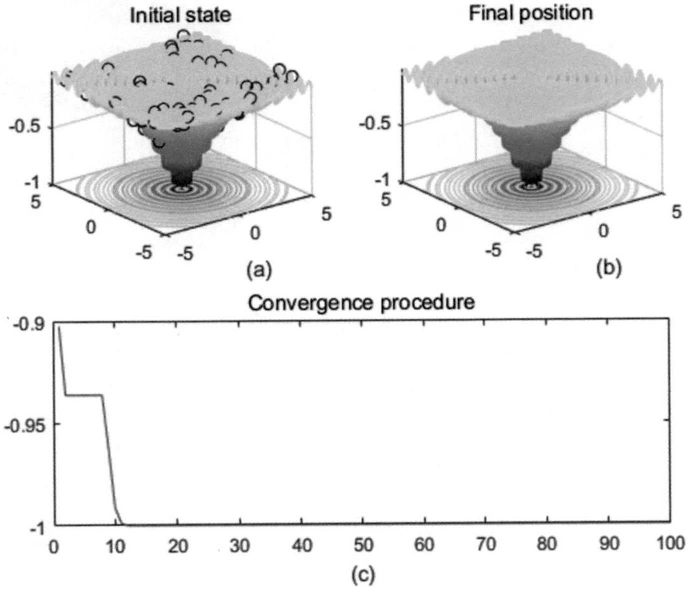

Fig. 5.3 Drop-wave

Table 5.2 Experimental parameter

Parameters	Value	Units	Parameters	Value	Units
ϑ_1	0.00431	day^{-1}	ϑ_2	1.02×10^{-9}	cell^{-1}
γ	6.41×10^{-11}	cell^{-1}	ε	0.08	day^{-1}
λ	0.204	day^{-1}	ξ_1	3.42×10^{-6}	cell^{-1}
ξ_2	2×10^{-11}	day^{-1}	α_1	0.0125	day^{-1}
α_2	0.125	day^{-1}	β_1	2.02×10^7	cell^2
β_2	2×10^7	cell	θ_1	0.1	day^{-1}
θ_2	1	day^{-1}	δ	0.95	N/A
ω	0.1392×10^{-4}	N/A			

According to clinical medical statistics borrowed from the literature [29], the specific parameters of the dynamic models are presented as Table 5.2. Based on the above, we have completed the establishment of the EEDS model, and determined the specific value of the cost function according to clinical needs. At the same time, the feasibility and effectiveness of the NLHO algorithm are also verified on benchmarks. Apply NLHO to the model of EEDS to develop therapeutic strategies. The best processing strategy is obtained through experiments, which proves the effectiveness and feasibility of the algorithm. The cost function is designed to minimize the number of tumor cells, and also to use the smallest dose of chemotherapeutic drugs and immune drugs to achieve the least harm to the human body.

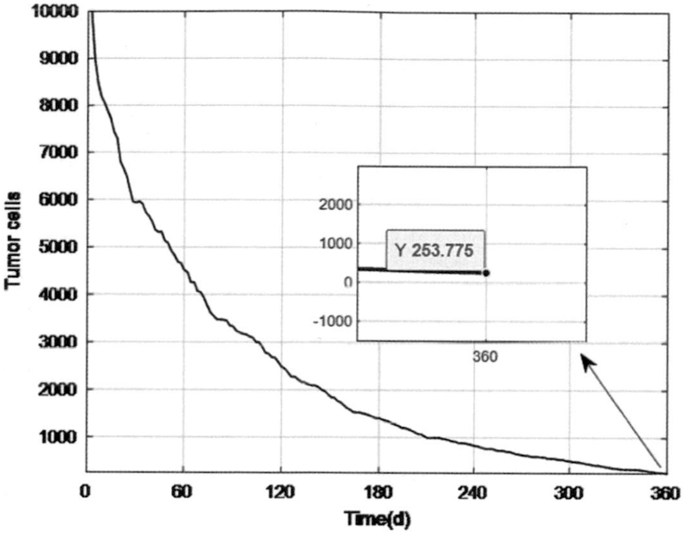

Fig. 5.4 Quantity curve of tumor cells

When we give the patient the initial number of tumor cells and immune cells, according to the EEDS model and follow certain chemotherapy and immunotherapy plans, we can get the following set of curves of tumor cells and immune cells. As shown in Figs. 5.4 and 5.5, it shows the quantity curve of tumor cells and immune cells. Within a one-year treatment period, the number of tumor cells was successfully reduced to 254. Although the number of immune cells was reduced to 1.52×106, significant effects were obtained for the treatment of tumors. Moreover, as shown in Figs. 5.7, 5.8 and 5.9, due to the decline of the body's immune cells, the immune drug dropped to 0.022 and then increased to about 0.03. This caused the concentration of immune drugs in human blood to rise from the trough of 0.0096 to 0.03, reaching a sufficient level.

The dosage of chemotherapeutic drugs is adaptively and dynamically changed, as shown in Fig. 5.6. Then the concentration of chemotherapeutic drugs in the blood also has the same changing trend, as shown in Fig. 5.8. The reason is to minimize the damage of drugs, and chemotherapy drugs are also harmful, not only killing tumor cells in the body but also destroying immune cells. If chemotherapeutic drugs are put in according to normal treatment methods, normal cells will suffer a lot of erosion, and the impact on the body will be even more significant. However, the drug dose optimized by NLHO will dynamically change adaptively, and the impact on normal cells will be appropriately reduced without affecting the elimination of tumor cells.

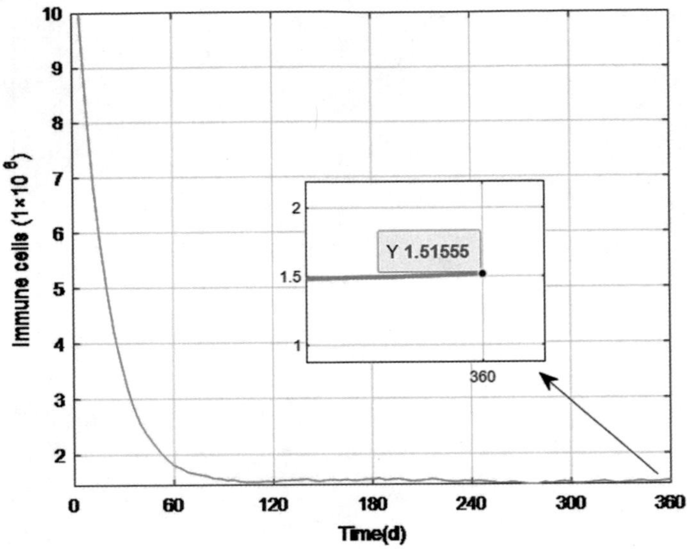

Fig. 5.5 Quantity curve of immune cells

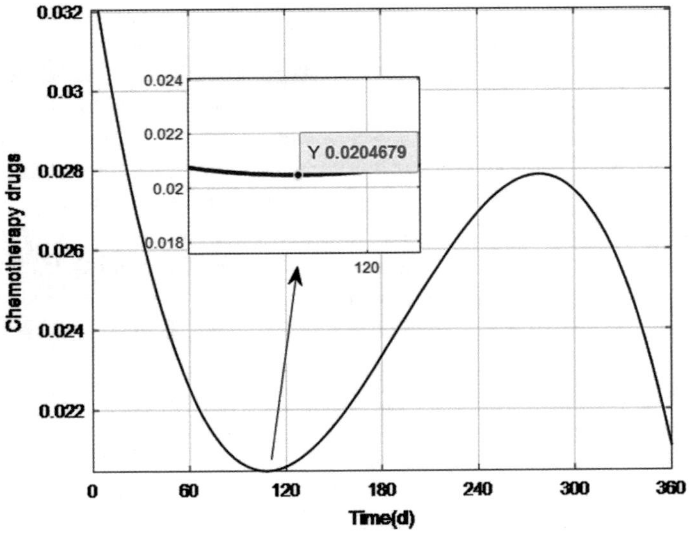

Fig. 5.6 Dosage of chemotherapeutic drugs

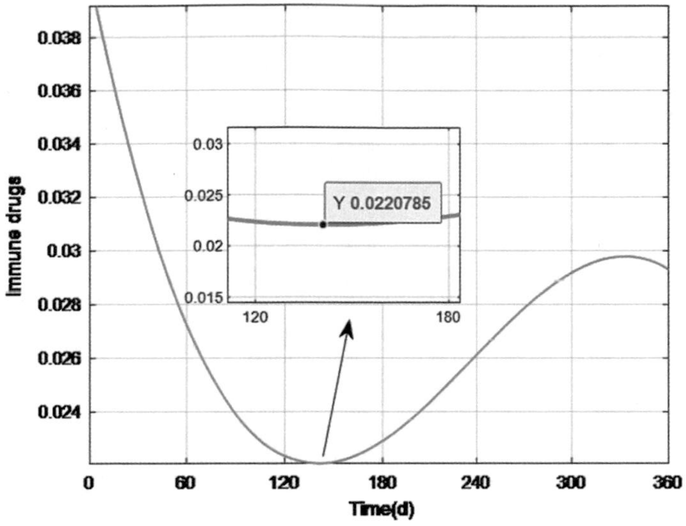

Fig. 5.7 Quantity curve of immune drug

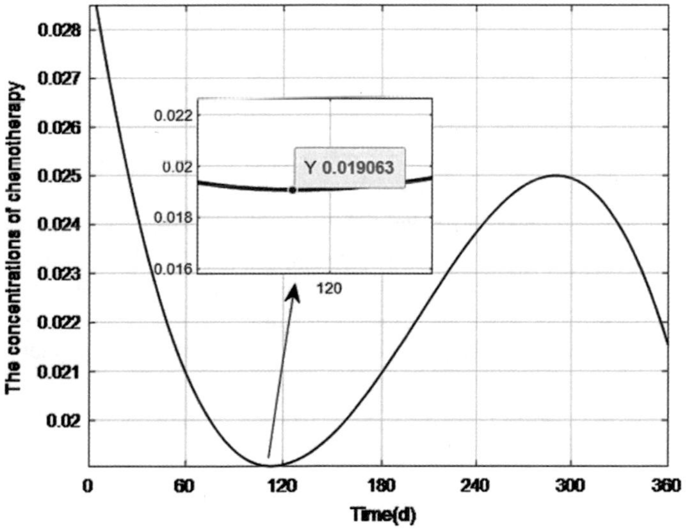

Fig. 5.8 Concentration of chemotherapy drugs

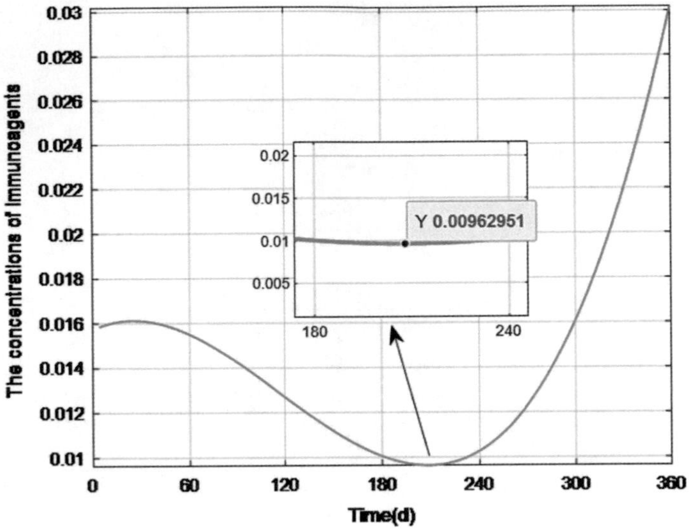

Fig. 5.9 Concentrations of immunereagents

5.6 Conclusion

It is a difficult problem to solve optimal therapeutic strategy for EEDS. Theoretically, it can achieve the desired therapeutic effect through reducing the tumor cells through the combined of chemotherapeutic drugs and immune drugs, and minimize the harm to the human body. Benefiting from the concept of heuristic algorithm in evolutionary computing, this chapter has designed the NLHO algorithm via 20 benchmark functions to test NLHO, including unimodal and multimodal, single-mode and multi-mode, single-extreme and multi-extreme, etc. It is compared with the five algorithms of TS, CSO, GA, ACO and SA, and runs independently 50 times to calculate the mean, minimum and standard deviation. It proves that NLHO has good optimization ability and can solve various problems well. At the same time, the development therapeutic strategies of EEDS have achieved very good results. The experimental results have shown that the NLHO algorithm develops therapeutic strategies well, and provides valuable prior knowledge and scientific basis for clinical medicine. Future work will further improve the EEDS, and integrate the optimal control strategy and the evolutionary calculation method for the optimal treatment method.

References

1. Lodhi I, Ahmad I, Uneeb M, Liaquat M (2019) Nonlinear control for growth of cancerous tumor cells. IEEE Access 7:177628–177636
2. de Pillis LG, Gu W, Radunskaya AE (2006) Mixed immunotherapy and chemotherapy of tumors: modeling, applications and biological interpretations. J Theor Biol 238(4):841–862
3. Sun J, Zhang H, Yan Y, Xu S, Fan X (2023) Optimal regulation strategy for nonzero-sum games of the immune system using adaptive dynamic programming. IEEE Trans Cybernet 53(3):1475–1484
4. Sharma S, Samanta GP (2016) Analysis of the dynamics of a tumor-immune system with chemotherapy and immunotherapy and quadratic optimal control. Differ Equ Dyn Syst 24(2):149–171
5. Sun J, Zhang H, Xu S, Liu Y (2023) Full information control for switched neural networks subject to fault and disturbance. IEEE Trans Neural Netw Learn Syst 34(2):703–714
6. Ogunmadeji B, Yusuf TT (2018) Optimal control strategy for improved cancer biochemotherapy outcome. Int J Sci Eng Res 9(12):583–600
7. Yang CWC, Wang CW, Hong RL, Kuo SH (2017) Treatment outcomes of and prognostic factors for definitive radiotherapy with and without chemotherapy for Stage I/II nasal extranodal NK/T-cell lymphoma. J Radiat Res 58(1):114–122
8. Tse S, Liang Y, Leung K, Lee K, Mok TS (2007) A memetic algorithm for multiple-drug cancer chemotherapy schedule optimization. IEEE Trans Syst Man Cybernet Part B (Cybernet) 37(1):84–91
9. Jiang H, Duerstock BS, Wachs JP (2018) Variability analysis on gestures for people with quadriplegia. IEEE Trans Cybernet 48(1):346–356
10. Wang J (2017) Spatial targeting of tumor-associated macrophage and tumor cells with a designer nanocarrier for cancer chemo-immunotherapy. In: 2017 39th annual international conference of the IEEE engineering in medicine and biology society (EMBC), p 291
11. Wang J, Huang M, Chen S, Luo Y, Shen S, Du X (2021) Nanomedicine-mediated ubiquitination inhibition boosts antitumor immune response via activation of dendritic cells. Nano Res 14:3900–3906
12. Li Y, Fan Y, Li K, Liu W, Tong S (2022) Adaptive optimized backstepping control-based rl algorithm for stochastic nonlinear systems with state constraints and its application. IEEE Trans Cybernet 52(10):10542–10555
13. Liu J, Feng J, Xiao Q, Liu S, Yang F, Lu S (2021) Fault diagnosis of rod pump oil well based on support vector machine using preprocessed indicator diagram. In: 2021 IEEE 10th data driven control and learning systems conference (DDCLS), pp 120–126
14. Chen C, Li A, Sun P, Xu J, Du W, Zhang J,..., Jiang X (2020) Efficiently restoring the tumoricidal immunity against resistant malignancies via an immune nanomodulator. J Control Release 324(10):574–585
15. Sherbet GV (1982) The biology of tumour malignancy. Academic, London
16. Sun J, Zhang H, Wang Y, Sun S (2022) Fault-tolerant control for stochastic switched IT2 fuzzy uncertain time-delayed nonlinear systems. IEEE Trans Cybernet 52(2):1335–1346
17. Rahman A, Kriak J, Meyer R, Goldblatt S, Rahman F (2020) A machine learning based modeling of the cytokine storm as it relates to COVID–19 using a virtual clinical semantic network (vCSN). In: 2020 IEEE international conference on big data (big data), pp 3803–3810
18. Zaharoff DA (2018) Engineering opportunities in cancer immunotherapy: after decades of missteps and delays, a growing immune-oncology market and improved cancer treatment outcomes open new prospects for biomedical engineers and data scientists. IEEE Pulse 9(4):8–11
19. Kerbel RS, Bertolini F, Man S, Hicklin DA, Emmenegger U, Shaked Y (2006) Antiangiogenic drugs as broadly effective chemosensitizing agents. Angiogenesis, pp 195–212
20. Ogunmadeji B, Yusuf T (2018) Optimal control strategy for improved cancer biochemotherapy outcome. Int J Sci Eng Res 9(12):583–600

21. Pinho STRD, Bacelar FS, Andrade RFS, Freedman HI (2013) A mathematical model for the effect of anti-angiogenic therapy in the treatment of cancer tumours by chemotherapy. Nonlinear Anal Real World Appl 14(1):815–828
22. Feng J, Liu J, Zhang H (2021) Speed control of pipeline inner detector based on interval dynamic matrix control with additional margin. IEEE Trans Ind Electron 68(12):12657–12667
23. Villasana M, Ochoa G (2004) Heuristic design of cancer chemotherapies. IEEE Trans Evol Comput 8(6):513–521
24. Liu J, Feng J, Gao X (2019) Fault diagnosis of rod pumping wells based on support vector machine optimized by improved chicken swarm optimization. IEEE Access 7:171598–171608
25. Liang Y, Leung KS, Mok TSK (2006) A novel evolutionary drug scheduling model in cancer chemotherapy. IEEE Trans Inf Technol Biomed 10(2):237–245
26. Vimalajeewa D, Balasubramaniam S, O'Brien B, Kulatunga C, Berry DP (2019) Leveraging social network analysis for characterizing cohesion of human-managed animals. IEEE Trans Comput Soc Syst 6(2):323–337
27. Li Y, Liu Y, Tong S (2022) Observer-based neuro-adaptive optimized control of strict-feedback nonlinear systems with state constraints. IEEE Trans Neural Netw Learn Syst 33(7):3131–3145
28. Yang Q, Chen WN, Gu T, Zhang H, Yuan H, Kwong S, Zhang J (2020) A distributed swarm optimizer with adaptive communication for large-scale optimization. IEEE Trans Cybernet 50(7):3393–3408
29. Sharma S, Samanta GP (2016) Analysis of the dynamics of a tumor-immune system with chemotherapy and immunotherapy and quadratic optimal control. Differ Equ Dyn Syst 24(2):149–171

Chapter 6
Combination Therapy-Based Adaptive Control for Organism Using Medicine Dosage Regulation Mechanism

6.1 Introduction

The death toll is soaring caused by neoplastic diseases, and the issues on nonlinear dynamics and control of tumour growth have motivated a widespread concern as [1]. Essential nutrients in humans are the resources for which the normal cells and tumor cells compete. Tumour cells will keep proliferating, robbing the limited energy supply of the body, and eventually disintegrating the somatic function to death. Somatic cells constantly divide, and new cells differentiate which end with apoptosis. In this manner, the relative balance can be maintained in human bodies. Nevertheless, when the process of differentiation for normal cells is out of control, the cells may well evolve into tumor cells. It is the nature for the tumor cells of which the tendency is to eat the body's nutrients crazily.

The population of tumour cells progressively increases for the following three characteristics. Firstly, the most obvious characteristic is the insensitivity to anti-growth signals. There exists strict control mechanism for normal cells, but for tumor cells, this mechanism is no longer valid. During the continuous process of division, tumor cells can escape from the monitoring of the anti-growth signals, which leads to the crazy growth of tumour cells. Secondly, tumour cells have the ability to promote the growth of blood vessels which are essential for providing nutrients, and it is the reason why the blood vessel density is associated with the malignant degree of the tumor tissue. Finally, tumor cells are also duplicitous, evolving camouflage abilities during their constant battle with immune cells to mislead the immune system into regarding them as normal cells, which results in the tumor immune escape. Thus, to suppress the growth of tumour cells, obstructing the generative mechanisms which relies on the necessary nutrients was an effective approach as literatures [2, 3].

Distinguishing from the mixed tumor treatment approach of immunotherapy and chemotherapy as [4], this chapter explores a more effective adaptive control strategy for organism using medicine dosage regulation mechanism. An additional population of cells which called endothelial cells enjoy the substances induced by malignant tumour cells, and they could transfer oxygen and nutrients to the primary focus caus-

J. Sun et al., *Adaptive Dynamic Programming*,
https://doi.org/10.1007/978-981-99-5929-7_6

ing proliferating of blood cells, which will increase carrying capacity of tumour cells known as tumour angiogenesis in [5]. As indicated in literature [6], anti-angiogenic agent could particularly decrease the growing rate of tumours, reaching saturation to some extent without killing the endothelial cells completely. When the chemotherapy agent was used in combination with anti-angiogenic agent to reduce the population of tumor cells , the latter could increase the effect of the former as described in [7]. Nevertheless, as the key element of promoting the growth of the vasculature, the endothelial cells could not be completely destroyed. Otherwise, it may not exist that the specified number of vasculature for constructing access of chemotherapy agent. On the basis of the pharmaceutical science concerning the chemotherapy agent and anti-angiogenic agent, the adaptive control strategy for organism will provide a guidance for clinical practice under the medicine dosage regulation mechanism, especially in the treatment process of Lung cancer. Furthermore, what counts is that since the anti-tumor drugs often kill both tumor cells and normal cells, it's of significance to utilize less drugs to achieve the therapeutic goal during the treatment process.

ADP is derived from dynamic programming and reinforcement learning, is a powerful tool to tackle optimization issues [8–10]. In general, the successful implementation of ADP-based methods depends on the cooperative work of actor and critic networks [11]. Under this framework, the actor is responsible for performing the control strategy with current data [12]. The goal of critic is to provide actor with the feedback information derived from the evaluation of the cost under the strategy. The distinct merit of this type of algorithm lies in that the optimal control strategy could be approximately acquired in the manner of iteration computation, and the "curse of dimensionality" could be obviated with effect. Different ADP-based methods have been researched by scholars to tackle multifarious optimal control problems with the aid of the artificial neural networks of which the performance is outstanding [13, 14], such as the robust control [15, 16], optimal consensus control [17–19] and the optimal tracking issues [20, 21]. Furthermore, for the system with multiple controllers, the optimization issues can be formulated by game theory. As a vital branch of game theory, NZSGs is derived from [22] with the goal of attaining the optimal strategy pair that can minimize the personal performance index for each player when stabilizing the controlled system [23–25]. Due to the excellent ability to approximate optimization, the ADP methods have been proposed to solve NZSGs. In [26], the adaptive method of critic-only structure was developed to solve two-player NZSGs without any initial stabilizing control. The experience replay technique was integrated into the ADP algorithm in [27] to concurrently utilize the historical data together with the real-time data to approximate the value function such that the persistence of excitation condition was not indispensable. In [28], the data-based integral reinforcement learning algorithm was proposed to solve NZSGs. More specially, it was a novel iterative learning algorithm based on both off-line and online manner which could extend the applicability of the data-based control scheme. Furthermore, in [29], the discrete-time N-player NZSGs was tackled via the off-policy reinforcement learning method which was independent of system dynamics.

Although the relevant academic achievements have been presented in theories and applications as [30–38], there is seldom any literature on this filed according to literature survey of the authors. The contributions can be shown as follows. First, the near-optimal therapy for the treatment of tumor is firstly acquired via the ADP approach which is an efficient adaptive intelligent learning algorithm. Second, the interactive system with discounted value function is constructed based on the mathematical model simulating the interaction relationships among cells and drugs. Besides, two kinds of chemotherapy drugs and a kind of anti-angiogenic agent participate in the therapy such that the combination therapeutic strategy can be derived under the architecture of NZSGs. Third, the idea of cybernetics is extended to the frontier fields of medicine, more precisely, the therapy of tumor. Under the MDRM, the derived therapeutic strategy can achieve the therapeutic goal with the lowest doses of drugs, and the practical indications for medicine is considered for the first time.

Notations: $\mathbb{N}^+$ denotes the set containing all positive integers. $\|\cdot\|$, $diag\{\cdot\}$ and $\nabla(\cdot) \triangleq \partial(\cdot)/\partial x$ respectively represent the Euclidean norm of a vector/matrix, the operation of constructing diagonal matrix and the gradient operator. $\lambda_m(\cdot)$ and $\lambda_M(\cdot)$ separately denote the minimum eigenvalue and maximum eigenvalue of a matrix. $I_{n\times n}$ is the unit matrix whose dimension is n.

6.2 Preliminaries

6.2.1 Establishment of Mathematical Model

In this section, the growth mathematical model is established which considers the interaction relationships among the normal cells, tumor cells and endothelial cells. Moreover, the effects of control inputs, i.e., the chemotherapy and anti-angiogenic drugs, on these cells are embodied in the model. Thus, in the model formed from ordinary differential equations as follows, $P_{NC}(t)$, $P_{TC}(t)$ and $P_{EC}(t)$ respectively represent the populations of normal cells, tumor cells and endothelial cells, $P_{CD_J}(t)(J = 1, 2)$ and $P_{AD}(t)$ denote the concentrations of chemotherapy and anti-angiogenic drugs.

The population of normal cells, which is influenced by tumor cells, endothelial cells and the concentrations of chemotherapy and anti-angiogenic drugs, is modeled by

$$\begin{aligned}\dot{P}_{NC}(t) =& \alpha_1 P_{NC}(t)\Big(1 - \frac{P_{NC}(t)}{C_1}\Big) - A_1 P_{NC}(t) P_{TC}(t)\\&- \Xi_1\big(P_{EC}(t), P_{AD}(t)\big)\frac{P_{NC}(t)P_{CD1}(t)}{B_1 + P_{NC}(t)}\\&- \Xi_2\big(P_{EC}(t), P_{AD}(t)\big)\frac{P_{NC}(t)P_{CD2}(t)}{B_1 + P_{NC}(t)},\end{aligned} \tag{6.1}$$

where $\Xi_\imath\big(P_{EC}(t), P_{AD}(t)\big) = \Xi_{\imath 1} P_{EC}(t) + \Xi_{\imath 2} P_{AD}(t) + \Xi_{\imath 0}$, $\imath = 1, 2$. The parameters α_1, B_1, C_1 denote the proliferation rate, Holling type 2 constant and carrying capacity for normal cells, respectively. A_1 is the contention parameter between normal cells and tumor cells.

As the tumor cells contend with normal cells for necessary nutrients, the population of tumor cells is affected by that of normal cells. Besides, there exist mutual effects among tumor cells, endothelial cells and the drugs. Thus the corresponding model can be written as

$$\begin{aligned}\dot{P}_{TC}(t) =& \alpha_2 P_{TC}(t) - \frac{\alpha_2 P_{TC}(t) P_{TC}(t)}{C_2 + \Phi P_{EC}(t)} - \Pi_1\big(P_{EC}(t), P_{AD}(t)\big)\frac{P_{TC}(t)P_{CD1}(t)}{B_2 + P_{TC}(t)} \\ &- \Pi_2\big(P_{EC}(t), P_{AD}(t)\big)\frac{P_{TC}(t)P_{CD2}(t)}{B_2 + P_{TC}(t)} - A_2 P_{NC}(t)P_{TC}(t),\end{aligned} \tag{6.2}$$

where $\Pi_j\big(P_{EC}(t), P_{AD}(t)\big) = \Pi_{j1}P_{EC}(t) + \Pi_{j2}P_{AD}(t) + \Pi_{j0}, \quad j = 1, 2.$ The parameters α_2, B_2, C_2 are multiplication rate, Holling type 2 constant and carrying capacity for tumor cells. A_2 is contention parameter between normal cells and tumor cells.

The population of endothelial cells is associated with tumor cells and anti-angiogenic drugs. The relations can be given as

$$\dot{P}_{EC}(t) = s_1 + K P_{TC}(t) + \alpha_3 P_{EC}(t)\Big(1 - \frac{P_{EC}(t)}{C_3}\Big) - \frac{\Xi_3 P_{EC}(t) P_{AD}(t)}{B_3 + P_{EC}(t)} \tag{6.3}$$

where K is multiplication rate caused by tumor cells and s_1 the inflow rate. Similarly, the parameters α_3, B_3, C_3 are multiplication rate, Holling type 2 constant and carrying capacity for endothelial cells. Ξ_3 is the killing rate for endothelial cells.

The concentrations of the drugs decrease during the treatment phases, owing to the washout process. Hence we can model the evolution process of the concentrations of chemotherapy and anti-angiogenic drugs by

$$\dot{P}_{CD1}(t) = Dr_{c1} - \Big(\beta_{c1} + m_1 \frac{P_{NC}(t)}{B_1 + P_{NC}(t)} + m_2 \frac{P_{TC}(t)}{B_2 + P_{TC}(t)}\Big) P_{CD1}(t) \tag{6.4}$$

$$\dot{P}_{CD2}(t) = Dr_{c2} - \Big(\beta_{c2} + m_3 \frac{P_{NC}(t)}{B_1 + P_{NC}(t)} + m_4 \frac{P_{TC}(t)}{B_2 + P_{TC}(t)}\Big) P_{CD2}(t) \tag{6.5}$$

and

$$\dot{P}_{AD}(t) = Dr_a - \Big(\beta_a + \frac{m_5 P_{EC}(t)}{B_3 + P_{EC}(t)}\Big) P_{AD}(t), \tag{6.6}$$

where Dr_{c1}, Dr_{c2} and Dr_a are the control inputs. β_{c1}, β_{c2} and β_a denote the washout rates for the drugs. m_1, m_2, m_3, m_4 and m_5 are the rates at which the drugs integrate into the cells. Based on the operations similar to that in [39], we obtain the simplified version of the model as

$$
\begin{cases}
\dot{p}_{NC}(t) = \alpha_1 p_{NC}(t)(1 - p_{NC}(t)) - a_1 p_{NC}(t) p_{TC}(t) \\
\qquad - \xi_1 \dfrac{p_{NC}(t) p_{CD1}(t)}{b_1 + p_{NC}(t)} - \xi_2 \dfrac{p_{NC}(t) p_{CD2}(t)}{b_1 + p_{NC}(t)}, \\
\dot{p}_{TC}(t) = \alpha_2 p_{TC}(t)\Big(1 - \dfrac{p_{TC}(t)}{1 + \phi p_{EC}(t)}\Big) - a_2 p_{NC}(t) p_{TC}(t) \\
\qquad - \pi_1 \dfrac{p_{TC}(t) p_{CD1}(t)}{b_2 + p_{TC}(t)} - \pi_2 \dfrac{p_{TC}(t) p_{CD2}(t)}{b_2 + p_{TC}(t)} \\
\dot{p}_{EC}(t) = s_1 + k p_{TC}(t) + \alpha_3 p_{EC}(t)(1 - p_{EC}(t)) - \xi_3 \dfrac{p_{EC}(t) p_{AD}(t)}{b_3 + p_{EC}(t)}, \\
\dot{p}_{CD1}(t) = u_{c1} - \Big(\beta_{c1} + m_1 \dfrac{p_{NC}(t)}{b_1 + p_{NC}(t)} + m_2 \dfrac{p_{TC}(t)}{b_2 + p_{TC}(t)}\Big) p_{CD1}(t), \\
\dot{p}_{CD2}(t) = u_{c2} - \Big(\beta_{c2} + m_3 \dfrac{p_{NC}(t)}{b_1 + p_{NC}(t)} + m_4 \dfrac{p_{TC}(t)}{B_2 + p_{TC}(t)}\Big) p_{CD2}(t), \\
\dot{p}_{AD}(t) = u_a - \Big(\beta_a + \dfrac{m_5 p_{EC}(t)}{b_3 + p_{EC}(t)}\Big) p_{AD}(t),
\end{cases}
\tag{6.7}
$$

where $\xi_\imath\big(p_{EC}(t), p_{AD}(t)\big) = \xi_{\imath 1} p_{EC}(t) + \xi_{\imath 2} p_{AD}(t) + \xi_{\imath 0}$ and $\pi_\jmath\big(p_{EC}(t), p_{AD}(t)\big) = \pi_{\jmath 1} p_{EC}(t) + \pi_{\jmath 2} p_{AD}(t) + \pi_{\jmath 0}$ with $\imath, \jmath = 1, 2$. The states $p_{NC}(t)$, $p_{TC}(t)$, $p_{EC}(t)$, $p_{CD1}(t)$, p_{CD2} and p_{AD} are nonnegative.

Remark 6.1 The differential equation (6.7) is the simplified model describing the interaction relationships among cells and drug. Observing the model, one can discover that there exists competition between normal cells and tumor cells. The tumor cells require more nutrients such that they facilitate the proliferation of endothelial cells, which could provide the indispensable nutrients to promote the growth of tumor. The tumor cells can be effectively damaged by the chemotherapy drugs which have side-effects on normal cells to some extent, and the anti-angiogenic drug contributes to the proliferation inhibition of the endothelial cells.

6.2.2 Nonzero-Sum Games Formulation

Consider the interaction model (6.7) rewritten as

$$
\begin{aligned}
\dot{x} &= f(x) + \begin{bmatrix} 0_{3\times 1} & 0_{3\times 1} \\ 0.06 & 0 \\ 0 & 0 \\ 0 & 0.12 \end{bmatrix} u_1 + \begin{bmatrix} 0_{3\times 1} & 0_{3\times 1} \\ 0 & 0 \\ 0.1 & 0 \\ 0 & 0 \end{bmatrix} u_2 \\
&= f(x) + g_1 u_1 + g_2 u_2,
\end{aligned}
\tag{6.8}
$$

where $u_1 = [u_{c1}, u_a]^T$, $u_2 = [u_{c2}, 0]^T$ and $f(x)$ is constructed by the right-hand side parts of (6.7) excluding the terms u_{c1}, u_{c2} and u_a.

Define the value function for player $\imath (\imath = 1, 2)$ as

$$V_\imath(x(t)) = \int_t^\infty e^{-\varrho_\imath(\varsigma - t)} \delta_\imath(x, u_1, u_2) d\varsigma, \tag{6.9}$$

where the utility function $\delta_\imath(x, u_1, u_2) = x^T \Upsilon_\imath x + u_1^T \mathcal{R}_{\imath 1} u_1 + u_2^T \mathcal{R}_{\imath 2} u_2$. The matrixes $\mathcal{R}_{\imath\jmath} (\imath, \jmath = 1, 2)$ and $\Upsilon_\imath$ are positive definite, and $\varrho_\imath > 0$ is the discount factor. According to the value function (6.9), we can define the corresponding Hamiltonian function as

$$\begin{aligned} H_\imath(x, u_1, u_2) = & (\nabla V_\imath)^T (f + g_1 u_1 + g_2 u_2) \\ & + \delta_\imath(x, u_1, u_2) - \varrho_\imath V_\imath, \imath = 1, 2. \end{aligned} \tag{6.10}$$

The optimal value function is defined as

$$V_\imath^* = \min_{u_\imath} \int_t^\infty e^{-\varrho_\imath(\varsigma - t)} (x^T \Upsilon_\imath x + \sum_{\jmath=1}^{N=2} u_\jmath^T \mathcal{R}_{\imath\jmath} u_\jmath) d\varsigma. \tag{6.11}$$

The target of NZSGs is to attain the admissible strategy pair $\{u_1^*, u_2^*\}$ with the definition given in [23, 40]. According to the stationarity condition, the optimal strategy for player $\imath$ could be obtained by

$$u_\imath^* = -\frac{1}{2} \mathcal{R}_{\imath\imath}^{-1} g_\imath^T \nabla V_\imath^*. \tag{6.12}$$

Thus the HJEs can be obtained as

$$\begin{aligned} H_\imath(x, u_1^*, u_2^*, \nabla V_\imath^*) = & (\nabla V_\imath^*)^T (f + g_1 u_1^* + g_2 u_2^*) \\ & + x^T \Upsilon_\imath x + u_1^* \mathcal{R}_{\imath 1} u_1^* + u_2^* \mathcal{R}_{\imath 2} u_2^* - \varrho V_\imath^* = 0. \end{aligned} \tag{6.13}$$

Remark 6.2 It's noteworthy that there exists no zero equilibrium for system (6.8), which may well result in the divergence of $V_\imath(x(t))$. To resolve this issue, the discounted factor $\varrho_\imath$ is introduced to form decay term such that $V_\imath(x(t))$ can be convergent.

In general, solving NZSGs is synonymous with solving the equations (6.13). Nevertheless, for nonlinear system, it's very intractable to tackle the coupled equations. To resolve this difficulty, an ADP method utilizing dosage regulation mechanism is proposed in the following sections.

6.3 MDRM-Based Adaptive Critic Learning Method for NZSGs

Firstly, we introduce the indications for medicine to judge when the medicine dosage should be regulated. Then under the MDRM, the ADP method of single-critic architecture is proposed to approximately seek the optimal strategy for the NZSGs of model (6.7).

6.3.1 MDRM-Based Optimal Strategy Derivation

For the sake of realizing conditioned therapy strategy, MDRM is required to handle the clinical data such that the strategy can be changed timely and necessarily. The time sequence $\{\hbar_\ell\}$ is constructed for recording the regulating instants and ℓ denotes the ℓth regulating instant. Then the state could be denoted as

$$\breve{x}_\ell(t) = x(\hbar_\ell), t \in [\hbar_\ell, \hbar_{\ell+1}). \tag{6.14}$$

For evaluating the difference between real-time data and latest recorded data, it's necessary to define an error function that $z_\ell = \breve{x}_\ell - x(t), t \in [\hbar_\ell, \hbar_{\ell+1})$. The operation of MDRM depends on the regulating condition which compares the error z_ℓ with the threshold associated with real-time data. The strategy is adjusted only when z_ℓ is larger than the threshold. That is, $\breve{u}_\imath = u_\imath(\breve{x}_\ell)$, $\imath = 1, 2$, and $\ell \in \mathbb{N}^+$. Thus the MDRM-based strategy could be got as

$$\breve{u}_\imath^* = -\frac{1}{2}\mathcal{R}_{\imath\imath}^{-1} g_\imath^T(\breve{x}_\ell)\nabla\breve{V}_\imath^*, \imath = 1, 2, \tag{6.15}$$

where $\nabla\breve{V}_\imath^* = \partial V_\imath^*/\partial x$ when $t = \hbar_\ell$. The version that based on the adjustment mechanism of HJEs is derived as

$$H_\imath(x, \breve{u}_1^*, \breve{u}_2^*, V_\imath^*) = \frac{1}{4}\sum_{\jmath=1}^{N=2}(\nabla\breve{V}_\jmath^*)^T g_\jmath(\breve{x}_\ell)\mathcal{R}_{\jmath\jmath}^{-1}\mathcal{R}_{\imath\jmath}\mathcal{R}_{\jmath\jmath}^{-1} g_\jmath^T(\breve{x}_\ell)\nabla\breve{V}_\jmath^*$$
$$+ (\nabla V_\imath^*)^T\left(f - \frac{1}{2}\sum_{\jmath=1}^{N=2} g_\jmath\mathcal{R}_{\jmath\jmath}^{-1} g_\jmath^T(\breve{x}_\ell)\nabla\breve{V}_\jmath^*\right) + x^T\Upsilon_\imath x - \varrho_\imath V_\imath^*. \tag{6.16}$$

Differing from HJEs (6.13), due to the existence of the error z_ℓ, (6.16) does not equal to zero. Before proceed with the discussion, the following assumption is required [41].

Assumption 6.1 The optimal strategy $u_\imath^*$ is locally Lipschitz. That is, for $\imath = 1, 2$, there exists a constant $\theta_\imath > 0$ such that $\|u_\imath^* - \breve{u}_\imath^*\|^2 \le \theta_\imath\|x - \breve{x}_\ell\|^2$.

Theorem 6.3 *Consider the system (6.8), and suppose that Assumption 6.1 holds and $V_\imath^*$ is the solution of (6.13). Then $\breve{u}_\imath^*$ formulated as (6.15) can stabilize system (6.8) when the following medicine indication is applied*

$$\|z_\ell\|^2 \leq \frac{(1-2\zeta)\lambda_m(\Upsilon)}{\theta\lambda_M(Y)}\|x\|^2, \tag{6.17}$$

where $\zeta \in (0, 1/2)$ is adjustable parameter. The terms θ, Υ and Y are given in (6.21) and (6.22).

Proof Selecting the Lyapunov function $L_{ya} = V_1^* + V_2^*$, we can obtain the corresponding derivative as

$$\dot{L}_{ya} = \sum_{\imath=1}^{N=2} (\nabla V_\imath^*)^T (f + g_1\breve{u}_1^* + g_2\breve{u}_2^*). \tag{6.18}$$

According to (6.13), we have

$$\begin{aligned}(\nabla V_\imath^*)^T f =& -(\nabla V_\imath^*)^T (g_1 u_1^* + g_2 u_2^*) - x^T \Upsilon_\imath x \\ &- u_1^{*T}\mathcal{R}_{\imath 1}u_1^* - u_2^{*T}\mathcal{R}_{\imath 2}u_2^* + \varrho_\imath V_\imath^*,\end{aligned} \tag{6.19}$$

and

$$(\nabla V_\imath^*)^T \sum_{\jmath=1}^{N=2} g_\jmath (u_\jmath^* - \breve{u}_\jmath^*) = -2u_\imath^{*T}\mathcal{R}_{\imath\imath}g_\imath^{-1}\sum_{\jmath=1}^{N=2} g_\jmath (u_\jmath^* - \breve{u}_\jmath^*). \tag{6.20}$$

Let $u^* = [u_1^{*T}, u_2^{*T}]^T$ and $\breve{u}^* = [(\breve{u}_1^* - u_1^*)^T, (\breve{u}_2^* - u_2^*)^T]^T$. Then we can derive that

$$\begin{aligned}\dot{L}_{ya} =& -x^T\Upsilon_1 x - x^T\Upsilon_2 x - \sum_{\imath=1}^{N=2}\sum_{\jmath=1}^{N=2} u_\jmath^{*T}\mathcal{R}_{\imath\jmath}u_\jmath^* \\ &+ 2\sum_{\imath=1}^{N=2} u_\imath^{*T}\mathcal{R}_{\imath\imath}g_\imath^{-1}\sum_{\jmath=1}^{N=2} g_\jmath(u_\jmath^* - \breve{u}_\jmath^*) + \varrho_1 V_1^* + \varrho_2 V_2^* \\ =& -x^T\Upsilon x - u^{*T}\mathcal{R}u^* - 2u^{*T}Z\breve{u}^* \\ &+ \varrho_1 V_1^* + \varrho_2 V_2^*,\end{aligned} \tag{6.21}$$

where $\Upsilon = \Upsilon_1 + \Upsilon_2$, $\mathcal{R} = diag\{\mathcal{R}_{11} + \mathcal{R}_{21}, \mathcal{R}_{12} + \mathcal{R}_{22}\}$, and $Z = [Z_1, Z_2]$ with $Z_\imath = [\mathcal{R}_{11}g_1^{-1}g_\imath, \mathcal{R}_{22}g_2^{-1}g_\imath]^T$, $\imath = 1, 2$. Applying Young's inequality, we have

$$\begin{aligned}\dot{L}_{ya} \leq & -x^T \Upsilon x - u^{*T} \mathcal{R} u^* + u^{*T} \mathcal{R} u^* \\ & + \breve{u}^{*T} Z^T \mathcal{R}^{-1} Z \breve{u}^* + \varrho_V \\ = & -x^T \Upsilon x + \breve{u}^{*T} Y \breve{u}^* + \varrho_V,\end{aligned} \tag{6.22}$$

where $Y = Z^T \mathcal{R}^{-1} Z$. It's noted that $u_\imath^*$ is the admissible strategy, we can derive that $V_\imath^*$ is bounded. Hence ϱ_V is the bound of the term $\varrho_1 V_1^* + \varrho_2 V_2^*$. According to the definitions of Υ and Y, we have that $\lambda_m(\Upsilon) > 0$ and $\lambda_M(Y) > 0$. Furthermore, we can obtain

$$\begin{aligned}\dot{L}_{ya} \leq & -2\zeta \lambda_m(\Upsilon) \|x\|^2 - (1 - 2\zeta) \lambda_m(\Upsilon) \|x\|^2 \\ & + \lambda_M(Y) \theta \|z_\ell\|^2 + \varrho_V,\end{aligned} \tag{6.23}$$

where $\theta = \theta_1 + \theta_2$. When the indication (6.17) is satisfied, we derive that $\dot{L}_{ya} \leq -2\zeta \lambda_m(\Upsilon) \|x\|^2 + \varrho_V$. Then we can find that $\dot{L}_{ya} < 0$ holds when $\|x\| > \sqrt{\frac{\varrho_V}{2\zeta \lambda_m(\Upsilon)}}$. In light of Lyapunov theorem, the strategy (6.15) can stabilize system (6.8). This completes the proof. ■

6.3.2 Implementation of Adaptive Critic Learning Method

In this section, the approximate optimal strategy under MDRM is derived by ADP method of single-critic architecture. In light of the universal approximation properties of neural networks (NNs), $V_\imath^*$ can be obtained by

$$V_\imath^* = \omega_\imath^{*T} \nu_\imath(x) + \sigma_\imath, \imath = 1, 2, \tag{6.24}$$

where $\omega_\imath^*$ is the ideal weight vector, $\nu_\imath$ the activation function and $\sigma_\imath$ the approximate error. To acquire the approximate version of the unknown vector $\omega_\imath^*$, the critic NN is constructed by

$$\hat{V}_\imath = \hat{\omega}_\imath^T \nu_\imath(x), \imath = 1, 2, \tag{6.25}$$

with $\hat{\omega}$ being the approximate vector. With the aid of critic NN, we can present the optimal strategy as

$$u_\imath^* = -\frac{1}{2} \mathcal{R}_{\imath\imath}^{-1} g_\imath^T \big((\nabla \nu_\imath)^T \omega_\imath^* + \nabla \sigma_\imath\big), \imath = 1, 2. \tag{6.26}$$

Accordingly, we can obtain the optimal and approximate optimal strategies under MDRM as

$$\breve{u}_\imath^* = -\frac{1}{2} \mathcal{R}_{\imath\imath}^{-1} g_\imath^T (\breve{x}_\ell) \big((\nabla \nu_\imath(\breve{x}_\ell))^T \omega_\imath^* + \nabla \sigma_\imath(\breve{x}_\ell)\big), \tag{6.27}$$

and

$$\breve{u}_\imath = -\frac{1}{2}\mathcal{R}_{\imath\imath}^{-1} g_\imath^T(\breve{x}_\ell)(\nabla\nu_\imath(\breve{x}_\ell))^T\hat{\omega}_\imath. \tag{6.28}$$

Then the approximate Hamiltonian can be presented as

$$H_\imath(x, \breve{u}_1, \breve{u}_2, \hat{\omega}_\imath) = \hat{\omega}_\imath^T\psi_\imath + \delta_\imath(x, \breve{u}_1, \breve{u}_2) \triangleq \epsilon_\imath, \tag{6.29}$$

where $\psi_\imath = \nabla\nu_\imath\big(f + g_1 u_1(\breve{x}_\ell) + g_2 u_2(\breve{x}_\ell)\big) - \varrho_\imath\nu_\imath$.

In order to minimize $\epsilon_\imath$ in (6.29), we set the target of minimization as $E = E_1 + E_2 = 1/2\epsilon_1^2 + 1/2\epsilon_2^2$. Via applying gradient descent approach, we obtain

$$\dot{\hat{\omega}}_\imath = -\gamma_\imath\frac{1}{(\psi_\imath^T\psi_\imath + 1)^2}\frac{\partial E}{\partial\hat{\omega}_\imath} = -\gamma_\imath\frac{\psi_\imath}{(\psi_\imath^T\psi_\imath + 1)^2}\epsilon_\imath = -\gamma_\imath\breve{\psi}_\imath\epsilon_\imath, \tag{6.30}$$

where $\gamma_\imath$ is the adjustable parameter and $\breve{\psi}_\imath = \psi_\imath/(\psi_\imath^T\psi_\imath + 1)^2$. Define $\tilde{\omega}_\imath = \omega_\imath^* - \hat{\omega}_\imath$. From (6.30), we derive that

$$\dot{\tilde{\omega}}_\imath = -\gamma_\imath\bar{\psi}_\imath\bar{\psi}_\imath^T\tilde{\omega}_\imath + \gamma_\imath\breve{\psi}_\imath e_\imath, \tag{6.31}$$

where $\bar{\psi}_\imath = \psi_\imath/(\psi_\imath^T\psi_\imath + 1)$ and the approximated residual error $e_\imath = -\nabla\sigma_\imath^T(f + g_1\breve{u}_1 + g_2\breve{u}_2) + \varrho_\imath\sigma_\imath$. For proceeding further, the following assumptions are required [11, 26, 27].

Assumption 6.2 For any $\imath \in \{1, 2\}$, the signal $\bar{\psi}_\imath$ is persistently excited on the time interval $[t, t + T]$. That is, there exists the positive constant $b_{\psi\imath}$ such that

$$b_{\psi\imath} I_{N_{c\imath}\times N_{c\imath}} \le \int_t^{t+T} \bar{\psi}_\imath\bar{\psi}_\imath^T d\varsigma, \tag{6.32}$$

with $N_{c\imath}$ being the neuron number of the $\imath$th critic network.

Assumption 6.3 For $\imath \in \{1, 2\}$, there exist positive constants such that $\|\omega_\imath^*\| \le b_{\omega\imath}$, $\|\nabla\nu_\imath\| \le b_{\nu\imath}$, $\|\nabla\sigma_\imath\| \le b_{\sigma\imath}$ and $\|e_\imath\| \le b_{e\imath}$.

6.4 Stability Analysis

In this section, the asymptotic stability of the controlled system is analyzed by applying Lyapunov theory. Before presenting the main results, the boundedness of critic weight is discussed in the following lemma.

Lemma 6.4 *For any $\imath \in \{1, 2\}$, suppose that Assumptions 6.2–6.3 hold and the initial weight is finite. If the critic tuning law (6.30) is applied, then it holds that $\tilde{\omega}_\imath$ is locally ultimately bounded.*

Proof Consider the Lyapunov function as $L_{y\omega}$. It's noted that the derivative of $\tilde{\omega}_\imath$ is flow dynamics, which indicates that there doesn't exist any jumps in the values of $\tilde{\omega}_\imath$. More specially, $\tilde{\omega}_\imath$ is continuous at the regulating instant. Thus we only need to consider the time interval between two adjoining regulating instants.

According to Assumptions 6.2–6.3, it can be derived that

$$\begin{aligned}\dot{L}_{y\omega} =& 2\gamma_1\tilde{\omega}_1^T\dot{\tilde{\omega}}_1 + 2\gamma_2\tilde{\omega}_2^T\dot{\tilde{\omega}}_2 \\ =& 2\gamma_1(-\tilde{\omega}_1\bar{\psi}_1\bar{\psi}_1^T\tilde{\omega}_1 + \tilde{\omega}_1\breve{\psi}_1 e_1) \\ &+ 2\gamma_2(-\tilde{\omega}_2\bar{\psi}_2\bar{\psi}_2^T\tilde{\omega}_2 + \tilde{\omega}_2\breve{\psi}_2 e_2).\end{aligned} \tag{6.33}$$

By applying Young's inequation, we can get

$$\begin{aligned}\dot{L}_{y\omega} \leq& -\gamma_1(\tilde{\omega}_1\bar{\psi}_1\bar{\psi}_1^T\tilde{\omega}_1 - e_1^T e_1) \\ &- \gamma_2(\tilde{\omega}_2\bar{\psi}_2\bar{\psi}_2^T\tilde{\omega}_2 - e_2^T e_2) \\ \leq& -\gamma_1 b_{\psi 1}\|\tilde{\omega}_1\|^2 - \gamma_2 b_{\psi 2}\|\tilde{\omega}_2\|^2 + \Gamma_1,\end{aligned} \tag{6.34}$$

where $\Gamma_1 = \gamma_1 b_{e1}^2 + \gamma_2 b_{e2}^2$. Furthermore, when $\|\tilde{\omega}_1\| > \sqrt{\frac{\Gamma_1}{\gamma_1 b_{\psi 1}}} \triangleq b_{\tilde{\omega}_1}$ or $\|\tilde{\omega}_2\| > \sqrt{\frac{\Gamma_1}{\gamma_2 b_{\psi 2}}} \triangleq b_{\tilde{\omega}_2}$, it yields that $\dot{L}_{y\omega} < 0$. The lemma is proved. ■

Theorem 6.4 *Consider the system (6.8) with strategy formulated as (6.28). Suppose that Assumptions 6.1–6.3 hold. The tuning law for critic network is given by (6.30). Then the state x and weight estimation error $\tilde{\omega}_\imath$ are UUB provided that the indication is applied*

$$\|z_\ell\|^2 \leq \frac{(1-\varpi_1^2)\lambda_m(\Upsilon)}{(1+\varpi_2)\theta\lambda_M(Y)}\|x\|^2 \triangleq \|z_e\|^2, \tag{6.35}$$

with ϖ_1 and ϖ_2 being the adjustable parameters.

Proof Select the Lyapunov function candidate as

$$\begin{aligned}L_Y =& \sum_{\imath=1}^{N=2} V_\imath^*(\breve{x}_\ell) + \sum_{\imath=1}^{N=2} V_\imath^*(x) + \frac{1}{2}\sum_{\imath=1}^{N=2}\tilde{\omega}_\imath^T\tilde{\omega}_\imath \\ =& L_{Ya} + L_{Yb} + L_{Yc}.\end{aligned} \tag{6.36}$$

Due to the utilization of MDRM, we present the proof process in two cases.

Case I: No regulation occurs, i.e., $t \in [\hbar_\ell, \hbar_{\ell+1})$. Then we obtain $\dot{L}_{Ya} = 0$. The derivative of L_{Yb} can be obtained as

$$\dot{L}_{Yb} = \sum_{\imath=1}^{N=2}(\nabla V_\imath^*)^T(f + g_1\breve{u}_1 + g_2\breve{u}_2). \tag{6.37}$$

Let $\breve{u} = [(\breve{u}_1 - u_1^*)^T, (\breve{u}_2 - u_2^*)^T]^T$. Applying the operations similar to that in Theorem 6.3, we have

$$\begin{aligned}\dot{L}_{Yb} \leq &- x^T \Upsilon x + \breve{u}^T Y \breve{u} + \varrho_V \\ \leq &- x^T \Upsilon x + \varrho_V + \lambda_M(Y)\|u_1^* - \breve{u}_1^* + \breve{u}_1^* - \breve{u}_1\|^2 \\ &+ \lambda_M(Y)\|u_2^* - \breve{u}_2^* + \breve{u}_2^* - \breve{u}_2\|^2 \\ \leq &- x^T \Upsilon x + \varrho_V + \lambda_M(Y)(1 + 1/\varpi_2)\|\breve{u}_1^* - \breve{u}_1\|^2 \\ &+ \lambda_M(Y)(1 + \varpi_2)\|u_1^* - \breve{u}_1^*\|^2 \\ &+ \lambda_M(Y)(1 + 1/\varpi_2)\|\breve{u}_2^* - \breve{u}_2\|^2 \\ &+ \lambda_M(Y)(1 + \varpi_2)\|u_2^* - \breve{u}_2^*\|^2. \end{aligned} \tag{6.38}$$

Recall that $\theta = \theta_1 + \theta_2$, and substitute (6.27) and (6.28) into (6.38). Then we can derive

$$\begin{aligned}\dot{L}_{Yb} \leq &- x^T \Upsilon x + (1 + \varpi_2)\theta\lambda_M(Y)\|x - \breve{x}_\ell\|^2 \\ &+ \varrho_V + \Gamma_2, \end{aligned} \tag{6.39}$$

where $\Gamma_2 = \frac{1}{4}\lambda_M(Y)(1 + 1/\varpi_2)^2\big(\|\mathcal{R}_{11}^{-1}\|^2 b_{g1}^2 b_{\nu1}^2 b_{\tilde{\omega}1}^2 + \|\mathcal{R}_{22}^{-1}\|^2 b_{g2}^2 b_{\nu2}^2 b_{\tilde{\omega}2}^2\big) + \frac{1}{4\varpi_2}$ $\lambda_M(Y)(1 + \varpi_2)^2\big(\|\mathcal{R}_{11}^{-1}\|^2 b_{g1}^2 b_{\sigma1}^2 + \|\mathcal{R}_{22}^{-1}\|^2 b_{g2}^2 b_{\sigma2}^2\big)$ with b_{g1} and b_{g2} denoting the bounds of known g_1 and g_2.

According to Assumption 6.2 and Assumption 6.3, we derive that

$$\dot{L}_{Yc} \leq -\gamma_1 b_{\psi1}\|\tilde{\omega}_1\|^2 - \gamma_2 b_{\psi2}\|\tilde{\omega}_2\|^2 + \Gamma_1. \tag{6.40}$$

Based on (6.39) and (6.40), we can obtain

$$\begin{aligned}\dot{L}_Y \leq &- (1 - \varpi_1^2)\lambda_m(\Upsilon)\|x\|^2 - \varpi_1^2\lambda_m(\Upsilon)\|x\|^2 \\ &+ (1 + \varpi_2)\lambda_M(Y)\theta\|x - \breve{x}_\ell\|^2 - \gamma_1 b_{\psi1}\|\tilde{\omega}_1\|^2 \\ &- \gamma_2 b_{\psi2}\|\tilde{\omega}_2\|^2 + £, \end{aligned} \tag{6.41}$$

where $£ = \Gamma_1 + \Gamma_2 + \varrho_V$. Applying the indication (6.35), then we conclude that $\dot{L}_Y < 0$ when one of the conditions hold that

$$\|x\| > \frac{1}{\varpi_1}\sqrt{\frac{£}{\lambda_m(\Upsilon)}} \triangleq \beta_x, \tag{6.42}$$

$$\|\tilde{\omega}_\iota\| > \sqrt{\frac{£}{\gamma_\iota b_{\psi\iota}}} \triangleq \beta_{\tilde{\omega}\iota}, \iota = 1, 2. \tag{6.43}$$

Thus x and $\tilde{\omega}_\imath$ can be guaranteed to be UUB.

Case II: A regulation occurs, that is, $t = \hbar_{\ell+1}$. The difference of L_Y can be given by

$$\triangle L_Y = \triangle L_{Ya} + \triangle L_{Yb} + \triangle L_{Yc}, \tag{6.44}$$

where the terms are defined by $\triangle L_{Ya} = V_1^*(\breve{x}_{\ell+1}) - V_1^*(\breve{x}_\ell) + V_2^*(\breve{x}_{\ell+1}) - V_2^*(\breve{x}_\ell)$, $\triangle L_{Yb} = V_1^*(x(\hbar_{\ell+1})) - V_1^*(x(\hbar_{\ell+1}^-)) + V_2^*(x(\hbar_{\ell+1})) - V_2^*(x(\hbar_{\ell+1}^-))$, $\triangle L_{Yc} = 1/2\tilde{\omega}_1^T(\hbar_{\ell+1})\tilde{\omega}_1(\hbar_{\ell+1}) - 1/2\tilde{\omega}_1^T(\hbar_{\ell+1}^-)\tilde{\omega}_1(\hbar_{\ell+1}^-) + 1/2\tilde{\omega}_2^T(\hbar_{\ell+1})\tilde{\omega}_2(\hbar_{\ell+1}) - 1/2\tilde{\omega}_2^T(\hbar_{\ell+1}^-)\tilde{\omega}_2(\hbar_{\ell+1}^-)$. Recalling the analysis in *Case I*, we obtain that $\dot{L}_Y < 0$ when x or $\tilde{\omega}_\imath$ is out of the corresponding bound. Furthermore, we can derive that $L_{Yb} + L_{Yc}$ is monotonically decreasing when $t \in [\hbar_\ell, \hbar_{\ell+1})$. In light of the properties of limits, we have

$$\begin{aligned} 0 \leq & V_\imath^*(x(\hbar_{\ell+1}^-)) + \frac{1}{2}\tilde{\omega}_\imath^T(\hbar_{\ell+1}^-)\tilde{\omega}_\imath(\hbar_{\ell+1}^-) \\ & - V_\imath^*(x(\hbar_{\ell+1})) - \frac{1}{2}\tilde{\omega}_\imath^T(\hbar_{\ell+1})\tilde{\omega}_\imath(\hbar_{\ell+1}). \end{aligned} \tag{6.45}$$

As x is proved to be UUB, we can obtain

$$V_\imath^*(\breve{x}_{\ell+1}) \leq V_\imath^*(\breve{x}_\ell). \tag{6.46}$$

According to (6.45) and (6.46), we can derive $\triangle L_Y < 0$, which indicates that the selected Lyapunov (6.36) is monotonically decreasing when $t = \hbar_{\ell+1}$. This completes the proof. ■

Remark 6.5 ϖ_1 and ϖ_2 in (6.35) are the adjustable parameters which determine the frequency of medicine dosage regulation. A larger ϖ_1 or ϖ_2 leads to a higher regulation frequency, and a smaller parameter implies a lower adjustment frequency. Thus we can determine these parameters according to the clinical data.

Remark 6.6 In thischapter, the approximate optimal combination therapeutic strategy is derived via ADP method to inhibit the proliferation of tumor cells under the mechanism of medicine dosage regulation. The MDRM is constructed on the foundation of the above-mentioned medicine indication (6.35). The data at the dosage-regulating instants should be recorded and will be utilized as reference data in the future. When the difference between the current clinical data and latest reference data is larger than the threshold, the medicine dosage can be regulated. Therefore, this mechanism can guarantee the derived therapeutic strategy to be regulated timely and necessarily.

Table 6.1 Parameter specifications of the cells

Parameters	Operations	Values	Parameters	Operations	Values
α_1	–	0.0068 day^{-1}	α_2	–	0.01 day^{-1}
α_3	–	0.002 day^{-1}	a_1	A_1C_1	0.00702 day^{-1}
a_2	A_2C_2	0.00072 day^{-1}	b_1	B_1/C_1	1.10
b_2	B_2/C_2	4.6205	b_3	B_3/C_3	4.6666
ϕ	$\Phi C_3/C_2$	0.1615	s_1	–	0.001
k	KC_2/C_3	0.00371day^{-1}			

Table 6.2 Parameter specifications of the drugs

Parameters	Operations	Values	Parameters	Operations	Values
ξ_{10}	Ξ_{10}/C_1	$1.2 \times 10^{-7}\ \text{day}^{-1}$	ξ_{11}	$\Xi_{11}C_3/C_1$	$4.2 \times 10^{-8}\ \text{day}^{-1}$
ξ_{12}	Ξ_{12}/C_1	$1.0 \times 10^{-7}\ \text{day}^{-1}$	π_{10}	Π_{10}/C_2	0.2051 day^{-1}
π_{11}	$\Pi_{11}C_3/C_2$	0.00431 day^{-1}	π_{12}	Π_{12}/C_2	19.4872 day^{-1}
ξ_{20}	Ξ_{20}/C_1	$6.0 \times 10^{-8}\ \text{day}^{-1}$	ξ_{21}	$\Xi_{21}C_3/C_1$	$2.2 \times 10^{-9}\ \text{day}^{-1}$
ξ_{22}	Ξ_{22}/C_1	$1.0 \times 10^{-8}\ \text{day}^{-1}$	π_{20}	Π_{20}/C_2	0.1251 day^{-1}
π_{21}	$\Pi_{21}C_3/C_2$	0.00217 day^{-1}	π_{22}	Π_{22}/C_2	15.7819 day^{-1}
ξ_3	Ξ_3/C_3	1.7143 day^{-1}	m_1	–	0.0002 day^{-1}
m_2	–	0.032 day^{-1}	m_3	–	0.0004 day^{-1}
m_4	–	0.028 day^{-1}	m_5	–	0.032 day^{-1}
β_{c1}	–	0.01813 day^{-1}	β_{c2}	–	0.01529 day^{-1}
β_a	–	0.136 day^{-1}			

6.5 Simulation Study

In this section, the mathematical model (6.7) is considered which presents the relations between cells and drugs. For simplicity, we have constructed the rephrased system (6.8) of which the control issue could be deemed as NZSGs.

In light of the clinical medical statistics and literature [38], the parameters on cells and drugs for model (6.7) are given in Table 6.1 and Table 6.2, respectively. For the discounted value function (6.9) of system (6.8), the corresponding parameters are set as $\mathcal{R}_{11} = 0.8I_{2\times2}$, $\mathcal{R}_{12} = 15I_{2\times2}$, $\mathcal{R}_{21} = 5I_{2\times2}$, $\mathcal{R}_{22} = I_{2\times2}$, $\Upsilon_1 = 0.02I_{6\times6}$ and $\Upsilon_2 = 0.06I_{6\times6}$. In addition, the discounted factors $\varrho_1 = \varrho_2 = 0.2$.

For the critic NNs, the activation functions are both set as $[x_1^2, x_1x_2, x_1x_3, x_1x_4, x_1x_5, x_1x_6, x_2^2, x_2x_3, x_2x_4, x_2x_5, x_2x_6, x_3^2, x_3x_4, x_3x_5, x_3x_6, x_4^2, x_4x_5, x_4x_6, x_5^2, x_5x_6, x_6^2]^T$, and the learning laws are set by $\gamma_1 = 1.5$ and $\gamma_2 = 2$. Besides, the parameters $\theta = 8$, $\varpi_1 = 0.8$ and $\varpi_2 = 8$.

The evolution curves of the model (6.7) are depicted in Fig. 6.1. From Fig. 6.1 we can observe that when $t = 200d$, the population of tumor cells reduces to zero, and when $t = 600d$, the population of normal cells almost returns to 1 and that of

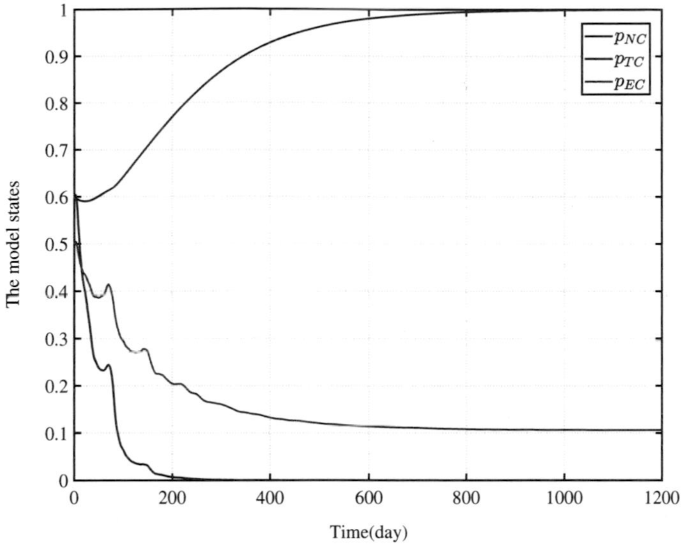

Fig. 6.1 The evolutions of model states

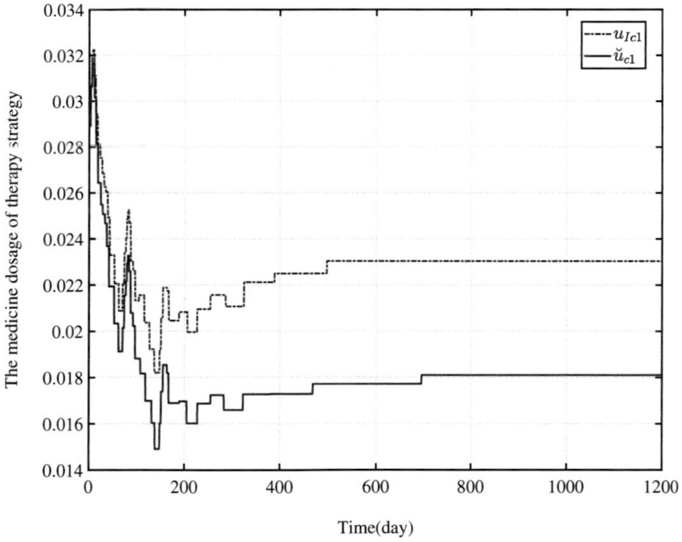

Fig. 6.2 The therapy strategy curves of chemotherapy drug 1

endothelial cells drops down to a small steady value. This indicates that the proliferation of tumor cells can be suppressed after 600 days under the optimal therapy strategy. In Figs. 6.1, 6.2, 6.3, 6.4, 6.5, 6.6 and 6.7, we compare the medicine dosages of the derived therapy strategy and that of initial therapy strategy. It indicates that the medicine dosages of our near-optimal therapy strategy are significantly less than the dosages of initial strategy. It's of great practical significance since superfluous drugs may well affect the health of patients and impose additional financial burdens

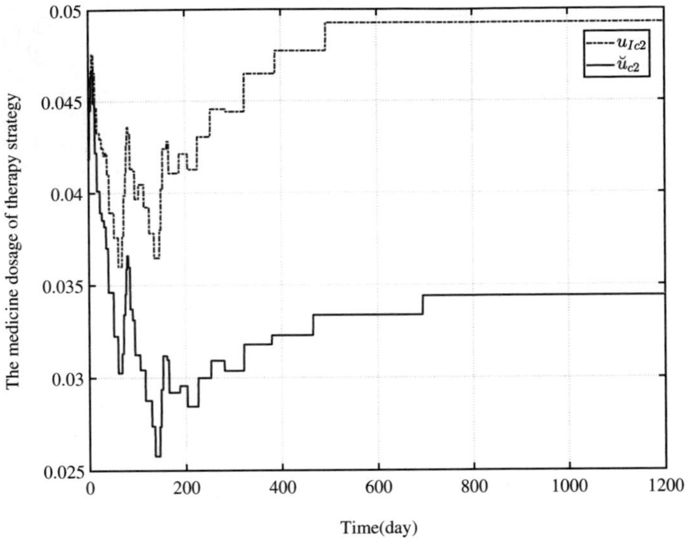

Fig. 6.3 The therapy strategy curves of chemotherapy drug 2

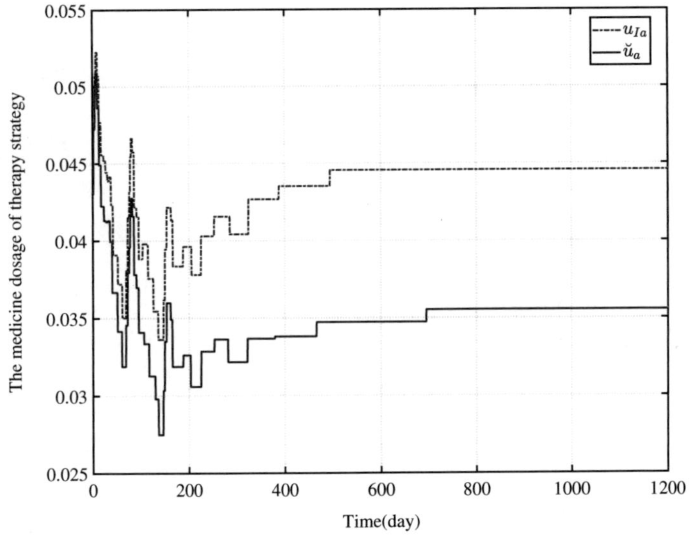

Fig. 6.4 The therapy strategy curves of anti-angiogenic drug

on patients. Besides, one can find that when the clinical data becomes better, the regulation frequency of the derived therapy strategy becomes lower. This implies that the therapy strategy based on medicine dosage regulation mechanism can be regulated with the indications for medicine timely and necessarily. Figures 6.5, 6.6 and 6.7 present the curves of the cells under different therapy strategies, that is,

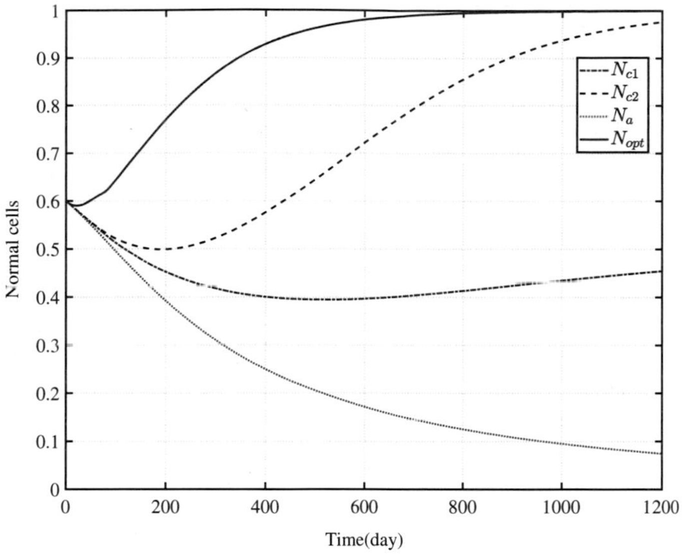

Fig. 6.5 The population of normal cells under different therapies

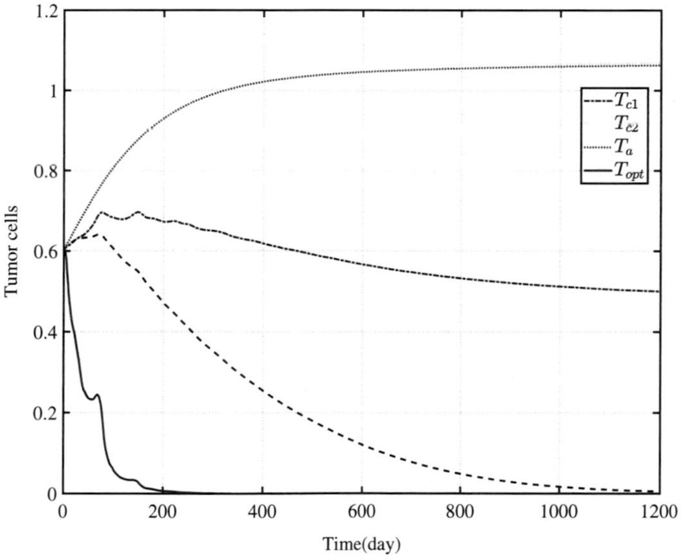

Fig. 6.6 The population of tumor cells under different therapies

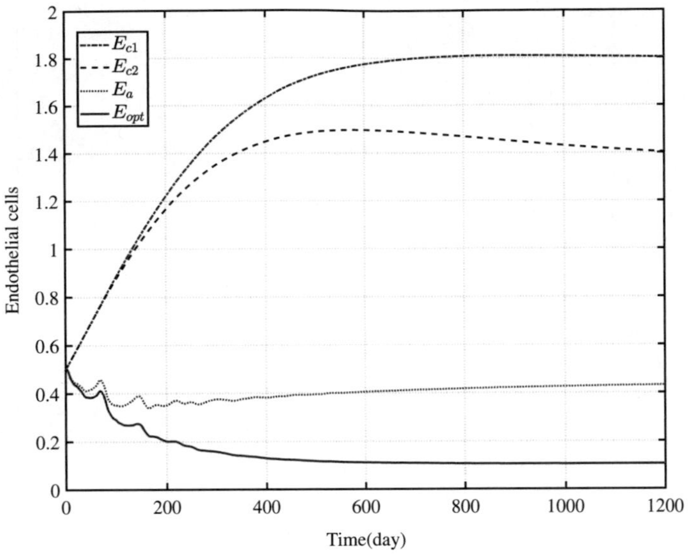

Fig. 6.7 The population of endothelial cells under different therapies

chemotherapy drug 1, chemotherapy drug 2, anti-angiogenic drug and the therapy comprised of these three drugs. We can conclude from Figs. 6.5, 6.6 and 6.7 that the therapeutic effect of the derived therapy is the best. Thus simulation results validate the effectiveness of our therapy strategy

6.6 Conclusion

In this chapter, an ADP-based method using medicine dosage regulation mechanism has been proposed to obtain the optimal combination therapy for curing cancer. A mathematical model is employed to describe the interactions among the normal cells, tumor cells, endothelial cells, chemotherapy drugs and anti-angiogenic drug. The mathematical model provides the foundation for us to solve the optimization issue under the architecture of NZSGs. The ADP method of single-critic framework is proposed to approximately seek the optimal strategy. In addition, the introduction of the medicine dosage adjustment mechanism guarantees the therapy strategy to be adjusted timely and necessary. Finally, the theory analysis and simulation results both indicate that the designed strategy can effectively decrease the population of tumor cells and endothelial cells with very few medicine dosage, which verifies the availability of the proposed method. Our future research direction is to seek the optimal strategy for decreasing tumor cells or other harmful cells with latest therapies, for example, the therapy applying oncolytic virus.

References

1. Sharma S, Samanta GP (2016) Analysis of the dynamics of a tumor-immune system with chemotherapy and immunotherapy and quadratic optimal control. Differ Equ Dyn Syst 24(2):149–171
2. Evans CM (1991) The metastatic cell, behaviour and biochemistry. Chapman and Hall, London
3. Sherbet GV (1982) The biology of tumour malignancy. Academic, London
4. de Pillis LG, Gu W, Radunskaya AE (2006) Mixed immunotherapy and chemotherapy of tumors: modeling, applications and biological interpretations. J Theor Biol 238(4):841–862
5. Bikfalvi A (1995) Significance of angiogenesis in tumour progression and metastass. Eur J Cancer 31(7–8):1101–1104
6. Beecken WC, Fernandes A, Joussen AM,..., Shing Y (2001) Effect of antiangiogenic therapy on slowly growing, poorly vascularized tumors in mice. J Natl Cancer Inst 93(5):382–387
7. Kerbel RS, Bertolini F, Man S, Hicklin DA, Emmenegger U, Shaked Y (2006) Antiangiogenic drugs as broadly effective chemosensitizing agents. Angiogenesis, pp 195–212
8. Harmon ME, Baird LC, Klopf AH (1995) Reinforcement learning applied to a differential game. Adapt Behav 4(1):3–28
9. Littman ML (2015) Reinforcement learning improves behaviour from evaluative feedback. Nature 521(7553):445–451
10. Li T, Yang D, Xie X, Zhang H (2022) Event-triggered control of nonlinear discrete-time system with unknown dynamics based on HDP(λ). IEEE Trans Cybernet 52(7):6046–6058
11. Vamvoudakis KG, Lewis FL (2010) Online actor-critic algorithm to solve the continuous-time infinite horizon optimal control problem. Automatica 46(5):878–888
12. Yang X, He H (2021) Decentralized event-triggered control for a class of nonlinear-interconnected systems using reinforcement learning. IEEE Trans Cybernet 51(2):635–648
13. Liu Y, Yao D, Li H, Lu R (2022) Distributed cooperative compound tracking control for a platoon of vehicles with adaptive NN. IEEE Trans Cybernet 52(7):7039–7048
14. Tan G, Wang Z, Shi Z (2023) Proportional-integral state estimator for quaternion-valued neural networks with time-varying delays. IEEE Trans Neural Netw Learn Syst 34(2):1074–1079
15. Liu D, Yang X, Wang D, Wei Q (2015) Reinforcement-learning-based robust controller design for continuous-time uncertain nonlinear systems subject to input constraints. IEEE Trans Cybernet 45(7):1372–1385
16. Wang D, Liu D, Li H, Luo B, Ma H (2016) An approximate optimal control approach for robust stabilization of a class of discrete-time nonlinear systems with uncertainties. IEEE Trans Syst Man Cybernet: Syst 6(5):713–717
17. Zhang H, Cai Y, Wang Y, Su H (2020) Adaptive bipartite event-triggered output consensus of heterogeneous linear multiagent systems under fixed and switching topologies. IEEE Trans Neural Netw Learn Syst 31(1):4816–4830
18. Wei Q, Liu D, Lewis FL (2015) Optimal distributed synchronization control for continuous-time heterogeneous multi-agent differential graphical games. Inf Sci 317:96–113
19. Zhang J, Zhang H, Feng T (2018) Distributed optimal consensus control for nonlinear multiagent system with unknown dynamic. IEEE Trans Neural Netw Learn Syst 29(8):3339–3348
20. Kamalapurkar R, Dinh H, Bhasin S, Dixon WE (2015) Approximate optimal trajectory tracking for continuous-time nonlinear systems. Automatica 51:40–48
21. Gao W, Jiang Z (2018) Learning-based adaptive optimal tracking control of strict-feedback nonlinear systems. IEEE Trans Neural Netw Learn Syst 29(6):2614–2624
22. Starr AW, Ho YC (1969) Nonzero-sum differential games. J Optim Theory Appl 3(3):184–206
23. Vamvoudakis KG, Lewis FL (2011) Online adaptive learning solution of coupled Hamilton-Jacobi equations for multi-player non-zero-sum games. Automatica 47(8):1556–1569
24. Zhu Y, Zhao D, Li X (2017) Iterative adaptive dynamic programming for solving unknown nonlinear zero-sum game based on online data. IEEE Trans Neural Netw Learn Syst 28(3):714–725

25. Song R, Li J, Lewis FL (2020) Robust optimal control for disturbed nonlinear zero-sum differential games based on single NN and least squares. IEEE Trans Syst Man Cybernet: Syst 50(11):4009–4019
26. Zhang H, Cui L, Luo Y (2013) Near-optimal control for nonzero-sum differential games of continuous-time nonlinear systems using single-network ADP. IEEE Trans Cybernet 43(1):206–216
27. Zhao D, Zhang Q, Wang D, Zhu Y (2016) Experience replay for optimal control of nonzero-sum game systems with unknown dynamics. IEEE Trans Cybernet 46(3):854–865
28. Zhang Q, Zhao D (2019) Data-based reinforcement learning for nonzero-sum games with unknown drift dynamics. IEEE Trans Cybernet 49(8):2874–2885
29. Song R, Wei Q, Zhang H, Lewis FL (2021) Discrete-time non-zero-sum games with completely unknown dynamics. IEEE Trans Cybernet 51(6):2929–2943
30. Zhang H, Qin C, Jiang B, Luo Y (2014) Online adaptive policy learning algorithm for H_∞ state feedback control of unknown affine nonlinear discrete-time systems. IEEE Trans Cybernet 44(12):2706–2718
31. Chen L, Zhu Y, Ahn CK (2023) Adaptive neural network-based observer design for switched systems with quantized measurements. IEEE Trans Neural Netw Learn Syst. https://doi.org/10.1109/TNNLS.2021.3131412
32. Zhang H, Su H, Zhang K, Luo Y (2019) Event-triggered adaptive dynamic programming for non-zero-sum games of unknown nonlinear systems via generalized fuzzy hyperbolic models. IEEE Trans Fuzzy Syst 27(11):2202–2214
33. Liu X, Ge SS, Zhao F, Mei X (2021) Optimized impedance adaptation of robot manipulator interacting with unknown environment. IEEE Trans Control Syst Technol 29(1):411–419
34. Massenio PR, Naso D, Lewis FL, Davoudi A (2020) Assistive power buffer control via adaptive dynamic programming. IEEE Trans Energy Convers 35(3):1534–1546
35. Ghasempour T, Nicholson GL, Kirkwood D, Fujiyama T, Heydecker B (2020) Distributed approximate dynamic control for traffic management of busy railway networks. IEEE Trans Intell Transp Syst 21(9):3788–3798
36. Wei Q, Liao Z, Shi G (2021) Generalized actor-critic learning optimal control in smart home energy management. IEEE Trans Ind Inf 17(10):6614–6623
37. Zhao J, Wang T, Pedrycz W, Wang W (2021) Granular prediction and dynamic scheduling based on adaptive dynamic programming for the blast furnace gas system. IEEE Trans Cybernet 51(4):2201–2214
38. Davari M, Gao W, Jiang ZP, Lewis FL (2021) An optimal primary frequency control based on adaptive dynamic programming for islanded modernized microgrids. IEEE Trans Autom Sci Eng 18(3):1109–1121
39. Pinho STRD, Bacelar FS, Andrade RFS, Freedman HI (2013) A mathematical model for the effect of anti-angiogenic therapy in the treatment of cancer tumours by chemotherapy. Nonlinear Anal Real World Appl 14(1):815–828
40. Liu D, Li H, Wang D (2014) Online synchronous approximate optimal learning algorithm for multiplayer nonzero-sum games with unknown dynamics. IEEE Trans Syst Man Cybernet: Syst 44(8):1015–1027
41. Yang X, Wei Q (2021) Adaptive critic learning for constrained optimal event-triggered control with discounted cost. IEEE Trans Neural Netw Learn Syst 32(1):91–104

Chapter 7
Adaptive Virotherapy Strategy for Organism with Constrained Input Using Medicine Dosage Regulation Mechanism

7.1 Introduction

Low efficacy and high toxicity for patients is the characteristics of traditional therapies as surgery, chemotherapy, and radiation, hence the most prosperous tumor treatment strategy, oncolytic virotherapy which depends on the virus with relatively weak pathogenicity and appropriate gene modification, simultaneously, the therapeutic effect benefits from strong replication capabilities. Similar to the principle of targeted therapy, gene-modified viruses repressed selectively infect tumor cells (ITCs) through rapid replication increment, and ultimately destroy TCs, concurrently, activate the body's immune response. Soluble tumor virus therapy not only can kill TCs, but also attract more immune cells to kill residual cancer cells, however, it doesn't deplete normal cells in the body. Oncolytic virus (OVs) enjoyed the superiority of minimal side effects and optimal therapeutic effects compared with traditional treatment strategies as literature [1]. Development of oncolytic viruses benefit from the virus-specific lytic CTL response eliciting immunostimulatory signals and contributing to killing of ITCs as literature [2], thus, viral doses, number of doses and timing with reliable mathematical models are the future research direction.

To lucubrate cancer virotherapy, mathematical models which described mechanisms of TCs, OVs and immune cells have been proposed and updated as literatures [3, 4]. Literature [5] expounded the inner mechanism including uninfected tumor cells (UTCs), ITCs and free viruses. Successively, the infected cells and uninfected cells are distinguished through logistic growth of TCs and elimination of free recombinant measles viruses as [6]. What matters most is the immune response which leads to inhibitory effect of viral therapy for misregarding of genetically modified viruses.

Therapy efficiency depends on hyperimmunity or not, in other words, infected cancer cells and viruses are swallowed for indistinguishability. Literatures has demonstrated the side effect of immune cells, and immunosuppressive agent cyclophosphamide is chosen to reduce immune response [7]. Reference [8] has considered the virus-free population adding the previous three variables, reflecting interactive relationship between innate immune with infected cancer cells and the virus cells,

J. Sun et al., *Adaptive Dynamic Programming*,
https://doi.org/10.1007/978-981-99-5929-7_7

evolving into an effective mechanism analysis model, but more effective control strategy is in urgent need. Cytokines form natural killer cells contribute most strength on destruction of both tumor and virus-infected cells. The proposed model gives explicitation of interplay among TCs, OVs, and immune response, which is the guideline of optimal therapeutic strategies or dosage regimen on oncotherapy. Although correlational research on regulation on immune system and TCs has been proposed using ADP as [9], selective oncolysis will enjoy optimal therapeutic effect through gene-modified viruses compared with wild-type OVs based on ADP method.

As a vital branch in machine learning, obtaining information from interactive environment [10–12], reinforcement learning (RL) has been demonstrated to perform well in solving optimal control issues of nonlinear systems [13]. The ADP method, which was derived from RL and dynamic programming, generally attempts to obtain the optimal strategies with the aid of the classic critic-actor algorithm framework [14]. Under this architecture, the critic evaluates the cost when the current strategy is applied, and actor updates the control strategy in accordance with the feedback information provided by the critic. Thus the approximate optimal strategy can be derived and the "curse of dimensionality" can be obviated. Recently, ADP-based methods have been widely researched to tackle various optimal issues, for instances, tracking control [15–17], optimal consensus control [18–20], zero-sum games and nonzero-sum games [21–23]. Different from fuzzy approximation as [24], the robust dynamic NN was established to asymptotically identify the uncertain system with additive disturbances, and the critic and actor worked together to find the equilibrium solution for nonzero-sum games subject to nonlinear system. The identifier was developed to reconstruct the unknown dynamics and the critic was tuned by a concurrent learning strategy which could effectively use real-time data and recorded data such that the persistence of excitation (PE) condition could be removed. By utilizing both online and off-line data, a data-based policy gradient ADP method was developed to seek optimal scheme in [25]. To address global optimum control issue and avoid falling into local optimality as[26], the ADP method which combined with the predesigned extra compensators was proposed in [27]. The introductions of these compensators contributed to deriving the augmented neighborhood error systems, thus the system dynamics requirement for ADP was avoided. In [28], integrating the neural network learning ability and the spirits of ADP, a general architecture of intelligent critic control was proposed to solve the robustness issues of disturbed nonlinear systems.

As saturation phenomena which exist widely in many practical systems can affect the system performance, multifarious ADP-based method were proposed to achieve optimal control with input constraints [29–31]. For the tumor-virus-immune system in this , the control input is the medicine containing the virus particles. Redundant or insufficient medicine dosages may well influence the therapeutic effect or patients' health. Thus we consider the asymmetric input constraints and construct the corresponding non-quadratic value functions associated with the tumor-virus-immune system.

Recently, ADP-based methods have been proposed to develop approximate optimal strategies in various practical applications [32–35]. However, there exist seldom any literatures associated with optimal strategy based on virotherapy which is derived from ADP-based methods. Enlightened by the literatures mentioned above, we design the virotherapy-based optimal strategy via ADP method with MDRM. The contributions can be stated as follows. Firstly, the mathematic model is introduced to simulate the relationships between TCs, OVs and immune cells. Due to the asymmetric dosage constraints for medicine, a non-quadratic utility function is constructed to form the discounted value function. Then, on the basis of the tumor-virus-immune model, ADP method of single-critic architecture is proposed to solve HJBE such that the approximate optimal strategy can be achieved, which means that the TCs can be largely eliminated with the constrained optimal virotherapy-based strategy. Furthermore, the reasonable the medicine dosage regulation mechanism is firstly introduced into this algorithm framework, and the indications for medicine is considered for the first time. Finally, theoretical analysis and simulation experiments both validate the effectiveness of the designed therapeutic strategy.

7.2 Problem Formulation and Preliminaries

7.2.1 Establishment of Interaction Model

In the section, tumor-virus-immune interaction model is introduced to describe the relations between TCs, viruses and immune cells. Due to the behavior of OVs, we can divide TCs into UTCs and ITCs. In the model composed of four ordinary differential equations as follows, $P_{TU}(t)$, $P_{TI}(t)$, $P_{VI}(t)$ and $P_{IM}(t)$ respectively denote the populations of UTCs, ITCs, free OVs and immune cells.

The population of UTCs can be affected by multiple factors, that is, the multiplication and apoptosis of TCs, the infection by OVs and the reduction caused by immune cells. Moreover, the growth dynamics of UTCs is presented as

$$\begin{aligned}\dot{P}_{TU}(t) = & A_1 P_{TU}(t)\big(1 - \frac{P_{TU}(t) + P_{TI}(t)}{K}\big) - A_2 P_{TU}(t) P_{VI}(t) \\ & - B_1 P_{TU}(t) P_{IM}(t) - C_1 P_{TU}(t),\end{aligned} \tag{7.1}$$

where A_1 is the tumor proliferation rate, A_2 is the infection rate of virus, B_1 denotes the killing-efficiency of immune cells, and C_1 is the apoptosis rate of UTCs.

Similarly, the population of ITCs can be modeled by

$$\dot{P}_{TI}(t) = A_2 P_{TU}(t) P_{VI}(t) - B_2 P_{TI}(t) P_{IM}(t) - \varphi P_{TI}(t), \tag{7.2}$$

where B_2 denotes the immune killing-efficiency of ITCs and φ is apoptosis rate of ITCs.

The lysis of ITCs which contain multiple replicated virion particles and the input of virus agentia can both contribute to the rise of the free virus population. Thus the evolution dynamics of virus population can be presented as

$$\begin{aligned}\dot{P}_{VI}(t) =& \mathcal{U} + \kappa\varphi P_{TI}(t) - A_2 P_{TU}(t) P_{VI}(t) \\ & - B_3 P_{VI}(t) P_{IM}(t) - C_2 P_{VI}(t),\end{aligned} \tag{7.3}$$

where $\mathcal{U}$ denotes the input of agentia, κ the burst size of free viruses, B_3 the immune killing-efficiency rate of OVs, and C_2 the clearance rate of OVs.

The immune response dynamics can be formulated as

$$\begin{aligned}\dot{P}_{IM}(t) =& D_1 P_{TI}(t) P_{IM}(t) + D_2 P_{TU}(t) P_{IM}(t) \\ & - C_3 P_{IM}(t),\end{aligned} \tag{7.4}$$

where D_1 and D_2 are immune response rates stimulated by infected and uninfected cells. And C_3 is the apoptosis rate of immune cells. For purpose of simplifying the interaction model, we utilize the nondimensionalization technique [36, 37] to derive the simplified version as

$$\begin{cases}\dot{p}_{TU}(t) = a_1 p_{TU}(t)(1 - p_{TU}(t) - p_{TI}(t)) - c_1 p_{TU}(t) \\ \qquad - a_2 p_{TU}(t) p_{VI}(t) - b_1 p_{TU}(t) p_{IM}(t) \\ \dot{p}_{TI}(t) = a_2 p_{TU}(t) p_{VI}(t) - b_2 p_{TI}(t) p_{IM}(t) - \varphi p_{TI}(t) \\ \dot{p}_{VI}(t) = u + \kappa p_{TI}(t) - a_2 p_{TU}(t) p_{VI}(t) \\ \qquad - b_3 p_{VI}(t) p_{IM}(t) - c_2 p_{VI}(t) \\ \dot{p}_{IM}(t) = d_1 p_{TI}(t) p_{IM}(t) + d_2 p_{TU}(t) p_{IM}(t) \\ \qquad - c_3 p_{IM}(t).\end{cases} \tag{7.5}$$

Herein the nonnegative states of nondimensionalization version are represented as $p_{TU}(t)$, $p_{TI}(t)$, $p_{VI}(t)$ and $p_{IM}(t)$.

Remark 7.1 In virotherapy, the viruses achieved their reproductive objective by infecting tumor cells and replicating themselves. After the lysis of infected cells, new reproductions burst out and infect other tumor cells. Under this mechanism, the tumor cells can be effectively eliminated. Furthermore, comparing with uninfected tumor cells, the infected cells can activate immune cells more effectually to kill tumor cells.

7.2.2 Problem Formulation

Consider the system (7.5) as

$$\dot{x} = f(x) + gu, \tag{7.6}$$

where $g = [0, 0, 1, 0]^T$, and $f(x)$ is constructed by the right-hand side parts of (7.5) excluding the control input u. $u \in [u_m, u_M]$ where u_m and u_M denote the minimum and maximum thresholds for medicine input dosage.

For system (7.6), the corresponding discounted value function is defined as

$$V(x(t)) = \int_t^{\infty} e^{-\theta(\iota - t)} W(x, u) d\iota, \tag{7.7}$$

with the discounted factor $\theta > 0$. The utility function is given by

$$W(x, u) = x^T \Upsilon x + \chi(u), \tag{7.8}$$

where the matrix Υ is positive definite, and $\chi(u)$ is non-negative function. It's noted that for system (7.6) the input constraints are not symmetric. In order to cope with this issue, function $\chi(u)$ is defined as

$$\chi(u) = 2\hbar \int_{\alpha}^{u} \psi^{-1}(\hbar^{-1}(\iota - \alpha)) d\iota, \tag{7.9}$$

where $\alpha = (u_m + u_M)/2$ and $\hbar = (u_M - u_m)/2$. $\psi(\cdot)$ is a monotonic odd function which is continuously differential with $\psi(0) = 0$. Without loss of generality, we select the hyperbolic tangent function as $\psi(\cdot)$, that is, $\psi(\cdot) = \tanh(\cdot)$.

Differentiating the value function (7.7) along system (7.6), we obtain that

$$0 = \nabla V^T (f + gu) + x^T \Upsilon x + \chi(u) - \theta V. \tag{7.10}$$

Then the Hamiltonian function can be expressed as

$$H(x, u, \nabla V) = \nabla V^T (f + gu) + x^T \Upsilon x + \chi(u) - \theta V. \tag{7.11}$$

The optimal value function is defined as

$$V^*(x) = \min_u \int_t^{\infty} e^{-\theta(\iota - t)} W(x, u) d\iota. \tag{7.12}$$

which satisfies HJBE

$$\min_u H(x, u, \nabla V^*) = 0. \tag{7.13}$$

Applying the stationary condition, we can derive the optimal strategy as

$$u^* = -\hbar \tanh(\frac{1}{2\hbar} g^T \nabla V^*) + \alpha. \tag{7.14}$$

On the basis of (7.13) and (7.14), we rewrite the HJBE as

$$
\begin{aligned}
(\nabla V^*)^T f - \hbar(\nabla V^*)^T g \tanh(\frac{1}{2\hbar} g^T \nabla V^*) + x^T \Upsilon x \\
+(\nabla V^*)^T g\alpha - \theta V^* + \chi(u^*) = 0.
\end{aligned} \tag{7.15}
$$

Remark 7.2 In the conventional optimal control issue with control constraints, it's often required that the input constraints should be symmetric. Nevertheless, the proposed method in this takes the asymmetric input constraints into account. Thus the symmetric constrained condition is relaxed by constructing the unconventional utility function (7.8).

Due to the nonlinear nature of (7.15), it's often intractable to derive the analytical solution, which is requisite for designing the optimal strategy. To overcome this issue, in the following sections, ADP method of single-critic network using dosage regulation mechanism is designed to approximately solve (7.15).

7.3 Optimal Strategy Based on MDRM

In order to achieve the goal of regulating therapeutic strategy timely and necessarily, MDRM is introduced to provide indications for medicine to determine the time when it's necessary to make some regulation. Therefore, the time sequence $\{z_\imath\}$ is required to record the regulating instants. The parameter $\imath \in \mathbb{N}^+$ represents the $\imath$th updating instant and $\mathbb{N}^+$ is the set including all positive integers. Then we can define the state as

$$
\breve{x}_\imath(t) = x(z_\imath), t \in [z_\imath, z_{\imath+1}). \tag{7.16}
$$

In general, the clinical data after the latest regulation is different from the current comparable data. Hence the error is given by

$$
\nu_\imath(t) = \breve{x}_\imath - x(t), t \in [z_\imath, z_{\imath+1}). \tag{7.17}
$$

Based on $\nu_\imath$ and the threshold associated with state x, the medicine regulation mechanism is established. When a regulation occurs, $\nu_\imath = 0$, which means the medicine dosage is regulated to be equal to the current medicine indication. The comparable data is updated by the clinical data at regulation instant, and the medicine dosage remains unchanged until the occurrence of the next regulation. That is, $\breve{u} = u(x_\imath)$. Thus we derive the MDRM-based strategy as

$$
\breve{u}^* = -\hbar \tanh(\frac{1}{2\hbar} g^T(\breve{x}_\imath) \nabla V^*(\breve{x}_\imath)) + \alpha, \tag{7.18}
$$

where $\nabla \breve{V}^* = \partial V^* / \partial x$ when $t = z_\imath$. Then the medicine regulation mechanism-based HJBE can be denoted as

$$
\begin{aligned}
H(x, \breve{u}^*, V^*) = & - \hbar(\nabla V^*)^T g \tanh(\frac{1}{2\hbar} g^T(\breve{x}_\iota)\nabla V^*(\breve{x}_\iota)) \\
& + (\nabla V^*)^T f + (\nabla V^*)^T g\alpha + x^T \Upsilon x \\
& + \chi(\breve{u}^*) - \theta V^*.
\end{aligned} \tag{7.19}
$$

The existence of the error ν_ι lead to that (7.19) does equal to 0, which is different from HJBE (7.15). Before proceeding, an assumption is necessary [31].

Assumption 7.1 The optimal strategy u^* is locally Lipschitz with respect to error ν_ι, i.e., $\|u^* - \breve{u}^*\|^2 \le K_u\|x - \breve{x}_\iota\|^2 = K_u\|\nu_\iota\|^2$ where K_u is a positive constant.

Theorem 7.1 *Consider the nonlinear system (7.6). Suppose that Assumption 7.1 is tenable and there exists function V^* satisfying (7.15). If the optimal strategy is formulated as (7.18) with the medicine indication*

$$
\|\nu_\iota\|^2 \le \frac{(1-\zeta^2)\lambda_m(\Upsilon)}{K_u}\|x\|^2 \tag{7.20}
$$

where $\zeta \in (0, 1)$ is the designed parameter, then the controlled system is guaranteed to be asymptotically stable in the sense of UUB.

Proof Select the Lyapunov function $\bar{Y} = V^*(x)$. Then we can obtain the derivative of V^*

$$
\dot{\bar{Y}} = (\nabla V^*)^T(f + g\breve{u}^*). \tag{7.21}
$$

According to (7.14) and (7.15), we derive that

$$
(\nabla V^*)^T f = -(\nabla V^*)^T g u^* - x^T \Upsilon x - \chi(u^*) + \theta V^*, \tag{7.22}
$$

and

$$
(\nabla V^*)^T g = -2\hbar(\tanh^{-1}((u^* - \alpha)/\hbar))^T. \tag{7.23}
$$

Then (7.21) can be rewritten as

$$
\begin{aligned}
\dot{\bar{Y}} = & - (\nabla V^*)^T g(u^* - \breve{u}^*) - x^T \Upsilon x - \chi(u^*) + \theta V^* \\
= & - 2\hbar(\tanh^{-1}((u^* - \alpha)/\hbar))^T(\breve{u}^* - u^*) - x^T \Upsilon x \\
& - \chi(u^*) + \theta V^* \\
= & - x^T \Upsilon x - \chi(u^*) + \theta V^* + \varpi,
\end{aligned} \tag{7.24}
$$

where $\varpi = -2\hbar(\tanh^{-1}((u^* - \alpha)/\hbar))^T(\breve{u}^* - u^*)$. Due to Young's inequality, from (7.24) we derive

$$
\varpi \le \hbar^2(\tanh^{-1}((u^* - \alpha)/\hbar))^2 + K_u\|\nu_\iota\|^2. \tag{7.25}
$$

Via variable substitution approach, we have

$$\chi(u^*) = 2\hbar \int_0^{u^*-\alpha} \tanh^{-1}((\iota - \alpha)/\hbar) d(\iota - \alpha). \tag{7.26}$$

The function (7.26) can be further expressed as

$$\begin{aligned}\chi(u^*) =& 2\hbar^2 \int_0^{\tanh^{-1}((u^*-\alpha)/\hbar)} \varsigma(1 - \tanh^2(\varsigma)) d\varsigma \\ =& -2\hbar^2 \int_0^{\tanh^{-1}((u^*-\alpha)/\hbar)} \varsigma \tanh^2(\varsigma) d\varsigma \\ &+ \hbar^2 (\tanh^{-1}((u^* - \alpha)/\hbar))^2.\end{aligned} \tag{7.27}$$

Based on (7.24), (7.25) and (7.27), we can obtain

$$\dot{\hat{Y}} \leq \Xi_1 + K_u \|\nu_\iota\|^2 + \theta V^* - x^T \Upsilon x, \tag{7.28}$$

where $\Xi_1(x) = 2\hbar^2 \int_0^{\tanh^{-1}((u^*-\alpha)/\hbar)} \varsigma \tanh^2(\varsigma) d\varsigma$. Via utilizing integral mean-value theorem, we derive that

$$\Xi_1(x) = 2\hbar^2 \tanh^{-1}((u^* - \alpha)/\hbar) \rho \tanh^2(\rho), \tag{7.29}$$

where $\rho \in (0, \tanh^{-1}((u^* - \alpha)/\hbar))$. As u^* is admissible, it can be deduced that V^* and ∇V^* are bounded. Let $\|V^*\| \leq b_V$ and $\|\nabla V^*\| \leq b_{\nabla V}$ with b_V and $b_{\nabla V}$ being positive constants. Then (7.29) becomes that

$$\begin{aligned}\Xi_1(x) \leq& 2\hbar^2 \tanh^{-1}((u^* - \alpha)/\hbar) \rho \\ \leq& 2\hbar^2 (\tanh^{-1}((u^* - \alpha)/\hbar))^2 \\ =& \frac{1}{2} \nabla V^{*T} g g^T \nabla V^* \\ =& \frac{1}{2} b_g^2 b_{\nabla V}^2 \triangleq b_{\Xi_1},\end{aligned} \tag{7.30}$$

where the positive constant b_g denotes the bound of $g(x)$. According to (7.28) and (7.30), it can be obtained that

$$\begin{aligned}\dot{\hat{Y}} \leq& -\zeta^2 \lambda_m(\Upsilon) \|x\|^2 - (1 - \zeta^2) \lambda_m(\Upsilon) \|x\|^2 \\ &+ K_u \|\nu_\iota\|^2 + \theta b_V + b_{\Xi_1}.\end{aligned} \tag{7.31}$$

When the indication (7.20) is satisfied, it yields that $\dot{\hat{Y}} \leq -\zeta^2 \lambda_m(\Upsilon) \|x\|^2 + \theta b_V + b_{\Xi_1}$. Then we can conclude that $\dot{\hat{Y}} < 0$ when $\|x\| > \sqrt{\frac{\theta b_V + b_{\Xi_1}}{\zeta^2 \lambda_m(\Upsilon)}}$. ■

Theorem 7.1 indicates that with the utilization of medicine regulation mechanism, the MDRM-based optimal strategy can asymptotically stabilize the controlled system.

7.4 MDRM-Based Approximate Optimal Control Design

The approximate optimal control strategy is designed based on the ADP algorithm which integrates the medicine regulation mechanism. Furthermore, for the closed-loop controlled system, the asymptotically stability in the sense of UUB is guaranteed when the proposed medicine indication is applied.

7.4.1 Implementation of the Adaptive Dynamic Programming Method

In this section, the approximate optimal strategy is designed by the ADP method of single-critic framework which integrates the medicine regulation mechanism.

Based on the universal approximation properties of NN, V^* can be represented as

$$V^* = \omega^{*T}\vartheta(x) + \tau, \tag{7.32}$$

where ω^* is the ideal weight vector, $\vartheta(\cdot)$ the activation function and τ the approximate error. Let $\Gamma_1(\breve{x}_\iota) = \frac{1}{2\hbar}g^T(\breve{x}_\iota)\nabla\vartheta^T(\breve{x}_\iota)\omega$, then we have

$$\breve{u}^* = -\hbar\tanh(\Gamma_1(\breve{x}_\iota)) + \bar{\tau}(\breve{x}_\iota) + \alpha, t \in [z_\iota, z_{\iota+1}) \tag{7.33}$$

where $\bar{\tau}(\breve{x}_\iota) = -(1/2)(1-\tanh^2(\Phi(\breve{x}_\iota)))g^T(\breve{x}_\iota)\nabla\tau(\breve{x}_\iota)$. Herein, $\Phi(\breve{x}_\iota)$ is selected between $1/(2\hbar)g^T(\breve{x}_\iota)\nabla V^*(\breve{x}_\iota)$ and $\Gamma_1(\breve{x}_\iota)$. As the ideal weight ω^* is unknown, the approximate version of V^* is derived by the critic NN, which is presented as

$$\hat{V} = \hat{\omega}^T\vartheta(x), \tag{7.34}$$

where $\hat{\omega}$ is the approximate vector. Then the MDRM-based approximate strategy can be obtained

$$\breve{u} = -\hbar\tanh(\Gamma_2(\breve{x}_\iota)) + \alpha, t \in [z_\iota, z_{\iota+1}), \tag{7.35}$$

where $\Gamma_2(\breve{x}_\iota) = 1/(2\hbar)g^T(\breve{x}_\iota)\nabla\vartheta^T(\breve{x}_\iota)\hat{\omega}$. Then the approximate Hamiltonian could be restated as

$$H(x, \breve{u}, \hat{\omega}) = \hat{\omega}^T \xi + x^T \Upsilon x + \chi(\breve{u}) \triangleq \varepsilon_H, \tag{7.36}$$

where $\xi = \nabla\vartheta(f + g\breve{u}) - \theta\vartheta$.

The goal of tuning $\hat{\omega}$ is to minimize the term ε_H. Thus we set the target function as $E = \frac{1}{2}\varepsilon_H^T \varepsilon_H$. Using the gradient descent approach, we obtain

$$\dot{\hat{\omega}} = -\ell \frac{\xi}{(\xi^T \xi + 1)^2} \varepsilon_H = -\ell \breve{\xi} \varepsilon_H, \tag{7.37}$$

where ℓ is the learning parameter and $\breve{\xi} = \xi/(\xi^T \xi + 1)^2$. Define $\tilde{\omega} = \omega^* - \hat{\omega}$. From (7.37) we derive that

$$\dot{\tilde{\omega}} = -\ell \bar{\xi} \bar{\xi}^T \tilde{\omega} + \ell \breve{\xi} e_H, \tag{7.38}$$

where $\bar{\xi} = \xi/(\xi^T \xi + 1)$ and the approximate residual error $e_H = -\nabla\tau^T(f + g\breve{u}) + \theta\tau$. Before presenting the main results, the following assumptions are requisite [38, 39].

Assumption 7.2 The signal $\bar{\xi}$ is persistently excited over the time interval $[t, t+T]$. In another word, there exists the positive constants ϕ and T such that

$$\phi I_{N_c \times N_c} \leq \int_t^{t+T} \bar{\xi}\bar{\xi}^T d\iota, \tag{7.39}$$

with N_c being the neuron number of the critic network.

Assumption 7.3 The terms $\bar{\tau}$ and e_H are both bounded. That is, $\|\bar{\tau}\| \leq b_{\bar{\tau}}$ and $\|e_H\| \leq b_{eH}$ where $b_{\bar{\tau}}$ and b_{eH} are positive constants.

7.4.2 Stability Analysis

This section discuss the asymptotic stability of the controlled system with the designed DARM-based strategy.

Theorem 7.2 *Consider system (7.6) and let Assumptions 7.1–7.3 hold. The strategy is given by (7.35) and the weights tuning law for critic is set as (7.37). Then the closed-loop system (7.6) and weight estimation error $\tilde{\omega}$ are asymptotically stable in the sense of UUB provided that the medicine indication is applied*

$$\|\nu_\iota\|^2 \leq \frac{(1 - \eta^2)\lambda_m(\Upsilon)}{2K_u} \|x\|^2 \triangleq \|T_{\nu_\iota}\| \tag{7.40}$$

with $\eta \in (0, 1)$ being the regulation parameter.

Proof Select the Lyapunov function as

$$Y = V^*(\breve{x}_\iota) + V^*(x) + \tilde{\omega}\ell^{-1}\tilde{\omega} = Y_a + Y_b + Y_c. \tag{7.41}$$

Note that when medicine indication is applied, the system can be described by the impulsive model comprising two components. One is flow dynamics for $t \in [z_\iota, z_{\iota+1})$ and the other is jump dynamics for $t = z_\iota$. Hence we present the discussions over the two cases.

Case I: No regulation occurs, i.e., $t \in [z_\iota, z_{\iota+1})$. Then we can obtain $\dot{Y}_a = 0$. In light of (7.22) and (7.23), we could derive that

$$\begin{aligned}\dot{Y}_b =&(\nabla V^*)^T(f + g\breve{u})\\ =&\Xi_2 - \chi(u^*) - x^T\Upsilon x + \theta V^*,\end{aligned} \tag{7.42}$$

where $\Xi_2 = -2\hbar(\tanh^{-1}((u^* - \alpha)/\hbar))^T(\breve{u} - u^*)$. According to Young's inequation, we have

$$\Xi_2 \le \hbar^2\|\tanh^{-1}((u^* - \alpha)/\hbar)\|^2 + \|\breve{u} - u^*\|^2. \tag{7.43}$$

Recalling (7.27), we obtain

$$\Xi_2 - \chi(u^*) \le \Xi_1(x) + \|\breve{u} - u^*\|^2. \tag{7.44}$$

As $\Xi_1(x)$ and $V^*(x)$ are bounded, (7.42) becomes

$$\dot{Y}_b \le \|\breve{u} - u^*\|^2 + b_{\Xi_1} + \theta b_V - x^T\Upsilon x. \tag{7.45}$$

Applying the Young's inequation, we derive that

$$\begin{aligned}\|\breve{u} - u^*\| =&\|\breve{u} - \breve{u}^* + \breve{u}^* - u^*\|^2 \le 2\|\breve{u} - \breve{u}^*\|^2 + 2\|\breve{u}^* - u^*\|^2\\ \le&4\|\hbar\tanh(\Gamma_1(\breve{x}_\iota)) - \hbar\tanh(\Gamma_2(\breve{x}_\iota))\|^2 + 4\|\bar{\tau}(\breve{x}_\iota)\|^2 + 2K_u\|\nu_\iota\|^2\\ \le&8\hbar^2\tanh^2(\Gamma_1(\breve{x}_\iota)) + 8\hbar^2\tanh^2(\Gamma_2(\breve{x}_\iota)) + 2K_u\|\nu_\iota\|^2 + 4b_{\bar{\tau}}^2.\end{aligned} \tag{7.46}$$

As $|\tanh(\cdot)| \le 1$, it could be obtained that

$$\dot{Y}_b \le -\lambda_m(\Upsilon)\|x\|^2 + 2K_u\|\nu_\iota\|^2 + \sigma, \tag{7.47}$$

where $\sigma = 16\hbar^2 + 4b_{\bar{\tau}}^2 + \theta b_V + b_{\Xi_1}$.

Taking the derivative of Y_c, we derive that

$$\dot{Y}_c = -2\tilde{\omega}^T\bar{\xi}\bar{\xi}^T\tilde{\omega} + 2\tilde{\omega}^T\breve{\xi}e_H. \tag{7.48}$$

In light of Young's inequation, it yields that

$$2\tilde{\omega}^T \check{\xi} e_H \leq 2\tilde{\omega}^T \bar{\xi} e_H \leq \tilde{\omega}^T \bar{\xi}\bar{\xi}^T \tilde{\omega} + e_H^T e_H. \tag{7.49}$$

Then (7.48) can be further expressed as

$$\dot{Y}_c \leq -\tilde{\omega}^T \bar{\xi}\bar{\xi}^T \tilde{\omega} + e_H^T e_H \leq -\lambda_m(\delta)\|\tilde{\omega}\|^2 + b_{eH}^2, \tag{7.50}$$

where $\delta = \bar{\xi}\bar{\xi}^T$.

According to (7.47) and (7.50), when the medicine indication (7.40) is satisfied, we can derive that

$$\begin{aligned}\dot{Y} \leq &-(1-\eta^2)\lambda_m(\Upsilon)\|x\|^2 - \eta^2\lambda_m(\Upsilon)\|x\|^2 + 2K_u\|\nu_\iota\|^2\\ &-\lambda_m(\delta)\|\tilde{\omega}\|^2 + b_{eH}^2 + \sigma\\ \leq &-\eta^2\lambda_m(\Upsilon)\|x\|^2 - \lambda_m(\delta)\|\tilde{\omega}\|^2 + b_{eH}^2 + \sigma.\end{aligned} \tag{7.51}$$

Then it can be concluded that $\dot{Y} < 0$ when one of the conditions holds that

$$\|x\| > \frac{1}{\eta}\sqrt{\frac{b_{eH}^2 + \sigma}{\lambda_m(\Upsilon)}}, \tag{7.52}$$

and

$$\|\tilde{\omega}\| > \sqrt{\frac{b_{eH}^2 + \sigma}{\lambda_m(\delta)}}. \tag{7.53}$$

Thus x and $\tilde{\omega}$ are demonstrated to be UUB.

Case II: A regulation occurs, i.e., $t = z_\iota$. The difference of L_Y is presented as

$$\begin{aligned}\triangle Y =& \underbrace{V^*(\check{x}_{\iota+1}) - V^*(\check{x}_\iota)}_{\triangle Y_a} + \underbrace{V^*(x(z_\iota^+)) - V^*(x(z_\iota))}_{\triangle Y_b}\\ =& \underbrace{\frac{1}{\ell}\tilde{\omega}^T(z_\iota^+)\tilde{\omega}(z_\iota^+) - \frac{1}{\ell}\tilde{\omega}^T(z_\iota)\tilde{\omega}(z_\iota)}_{\triangle Y_c}.\end{aligned} \tag{7.54}$$

From the analysis in *Case I*, it can be derived that $\dot{L}_Y < 0$ when (7.52) or (7.53) is satisfied. It can be further deduced that $Y_b + Y_c$ is monotonically decreasing when $t \in [z_\iota, z_{\iota+1})$, that is,

$$Y_b(x(z_\iota)) + Y_c(x(z_\iota)) \geq Y_b(x(z_\iota + \epsilon)) + Y_c(x(z_\iota + \epsilon)), \tag{7.55}$$

where $\epsilon \in (0, z_{\iota+1} - z_\iota)$. According to the properties of limits, we can obtain

$$Y_b(x(z_\iota)) + Y_c(x(z_\iota)) \geq Y_b(x(z_\iota^+)) + Y_c(x(z_\iota^+)), \tag{7.56}$$

with $x(z_\iota^+) = \lim_{\epsilon \to 0} x(z_\iota + \epsilon)$. More specially, it yields that

$$V^*(x(z_\iota)) + \frac{1}{\ell}\tilde{\omega}^T(z_\iota)\tilde{\omega}(z_\iota) \geq V^*(x(z_\iota^+)) + \frac{1}{\ell}\tilde{\omega}^T(z_\iota^+)\tilde{\omega}(z_\iota^+). \tag{7.57}$$

As x is proved to be UUB, it can be obtained that

$$V^*(\breve{x}_{\iota+1}) \leq V^*(\breve{x}_\iota). \tag{7.58}$$

From (7.57) and (7.58), it's derived that $\Delta Y < 0$, which indicates that the constructed Lyapunov (7.41) is monotonically decreasing when $t = z_\iota$. ■

Remark 7.3 ζ in (7.40) is the regulation parameter determining the frequency of medicine dosage regulation. A large ζ means that the medicine dosage is regulated frequently while a small ζ implies the regulation occurs rarely. It can be set as an appropriate value according to the clinical data.

Remark 7.4 Theorem 7.2 indicates that the designed MDRM-based approximate optimal strategy (7.35) can asymptotically stabilize system (7.6). The medicine indication (7.40), the cornerstone of MDRM, can provide a reasonable reference threshold for therapeutic strategy. When the difference derived from the current clinical data and latest reference data is larger than the threshold, the medicine dosage can be regulated, and the current indication data will be recorded and utilized as the new reference data in the future. Thus the designed therapeutic strategy can be regulated timely and necessarily according to the medicine indication.

Remark 7.5 The discount factor is programmed to avoid infinity and infinitesimal value function in the accumulation of rewards, and immediately return can earn more than the delayed return of interest. In human trials, we have found that human prefer to immediately return can present close to exponential growth, the discount factor is used to simulate such a cognitive model and biological process to make a decision.

7.5 Simulation Study

In this section, we consider the system (7.6) which is the simplified version of the growth dynamics of cells and viruses described by (7.1)–(7.4). Based on system (7.6), the simulation experiment is conducted to show the effectiveness of the proposed ADP method with medicine regulation mechanism.

According to the clinical medical statistics and literatures [36, 37, 40], the parameters associated with the dynamics (7.1)–(7.4) are presented in Table 7.1. After the nondimensionalization, the corresponding parameters are set as $a_1 = 0.36, a_2 = 0.1$, $b_1 = 0.36, b_2 = 0.48, b_3 = 0.16, c_1 = 0.1278, c_2 = 0.2, c_3 = 0.036, d_1 = 0.6$, and

Table 7.1 Parameter specifications of the tumor-virus-immune system

Parameters	Descriptions	Values
A_1	Tumor proliferation rate	$2 \times 10^{-2}\,\mathrm{h}^{-1}$
A_2	Infection rate of virus	$7 \times 10^{-10}\,\mathrm{mm}^3/\mathrm{h}$
B_1	Killing-efficiency of immune cells	$2 \times 10^{-8}\,\mathrm{mm}^3/\mathrm{h}$
B_2	Immune killing-efficiency of infected tumor cells	$2 \times 10^{-8}\,\mathrm{mm}^3/\mathrm{h}$
C_1	Apoptosis rate of uninfected tumor cells	$0.0071\,\mathrm{h}^{-1}$
C_2	Clearance rate of viruses	$0.0119\,\mathrm{h}^{-1}$
C_3	Apoptosis rate of immune cells	$0.002\,\mathrm{h}^{-1}$
D_1	Immune response rate stimulated by infected cells	$5.6 \times 10^{-7}\,\mathrm{mm}^3/\mathrm{h}$
D_2	Immune response rate stimulated by uninfected cells	$5.6 \times 10^{-7}\,\mathrm{mm}^3/\mathrm{h}$
φ	Apoptosis rate of infected tumor cells	$0.056\,\mathrm{h}^{-1}$
κ	Burst size of free virus	9.0

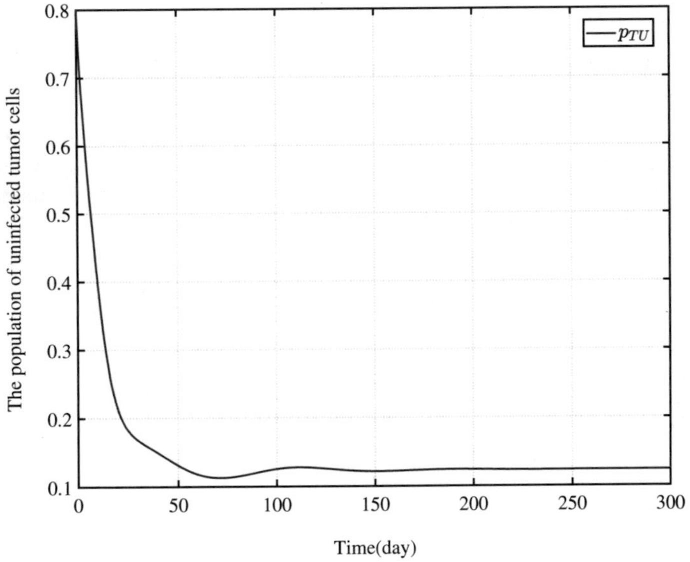

Fig. 7.1 The population evolution of uninfected tumor cells

$d_2 = 0.29$. The initial state vector is $[0.8, 0, 0.2, 0.05]^T$. The minimum and maximum thresholds are given by $u_m = 0$ and $u_M = 0.02$. For the discounted value function (7.7) of system (7.6), the parameters $\Upsilon = 0.2 I_{4\times 4}$ and $\theta = 0.5$.

For the critic network, we select the activation function as $[x_1^2, x_1x_2, x_1x_3, x_1x_4, x_2^2, x_2x_3, x_2x_4, x_3^2, x_3x_4, x_4^2]^T$. The other parameters are respectively set as $K_u = 20$, $\zeta = 0.9$ and $\ell = 1.6$.

Simulation results demonstrate that in Figs. 7.1, 7.2, 7.3, 7.4, 7.5, 7.6 and 7.7. For model (7.5), the evolution trajectories of states are respectively depicted in

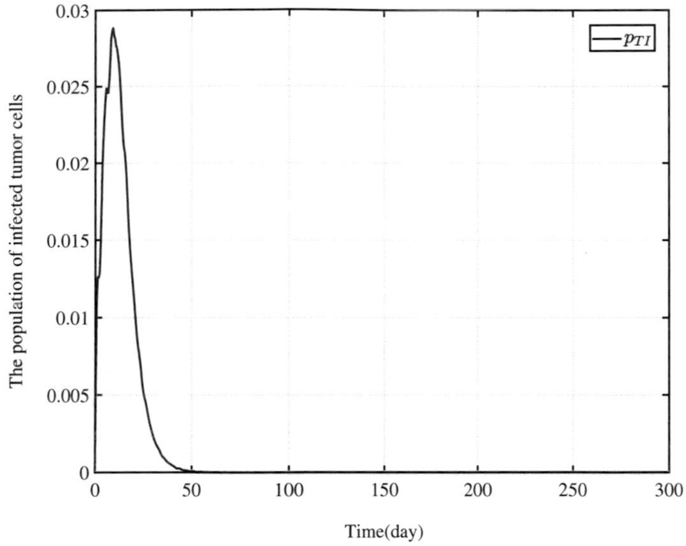

Fig. 7.2 The population evolution of infected tumor cells

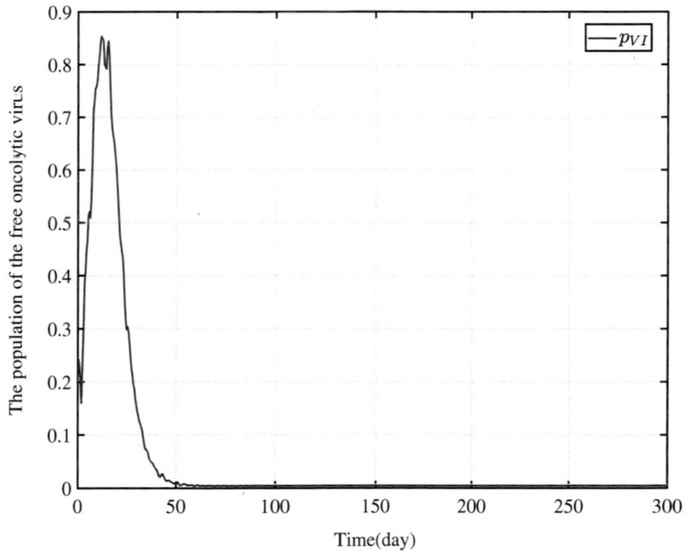

Fig. 7.3 The population evolution of free oncolytic virus

Figs. 7.1, 7.2, 7.3 and 7.4. From Fig. 7.1, we could observe that under the attacks from oncolytic viruses and immune cells, the population of uninfected tumor cells rapidly declines and reaches a stabilizing value which is very low after $t = 150d$. Figures 7.2 and 7.3 reveal the relations between the population of infected tumor cells

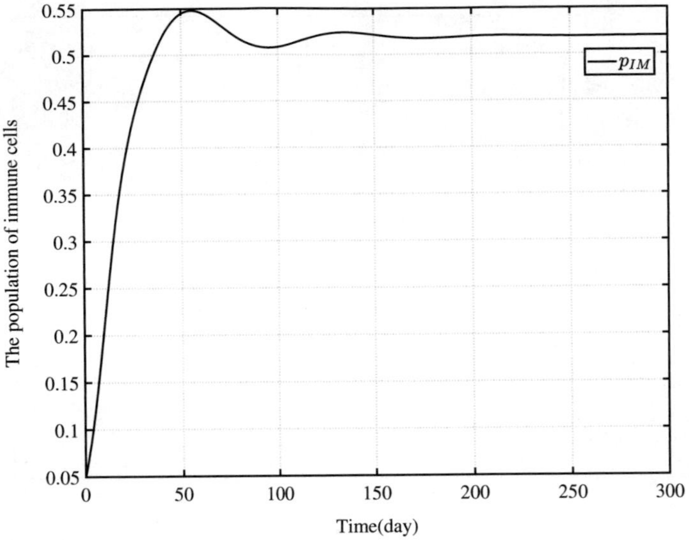

Fig. 7.4 The population evolution of immune cells

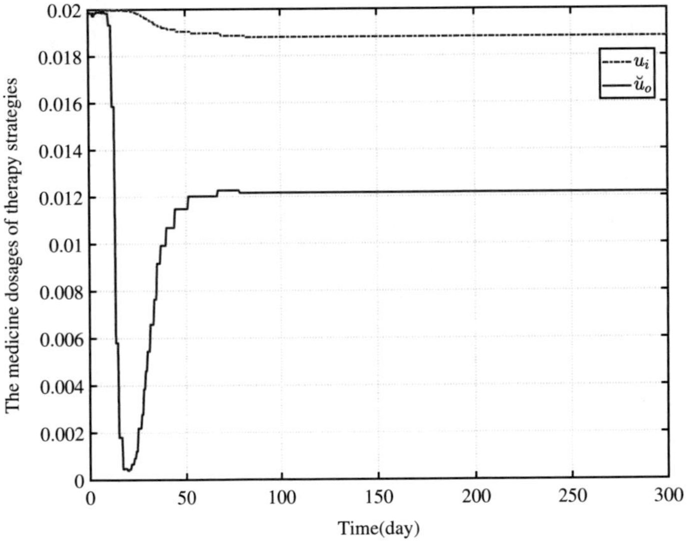

Fig. 7.5 The curves of the therapeutic strategies

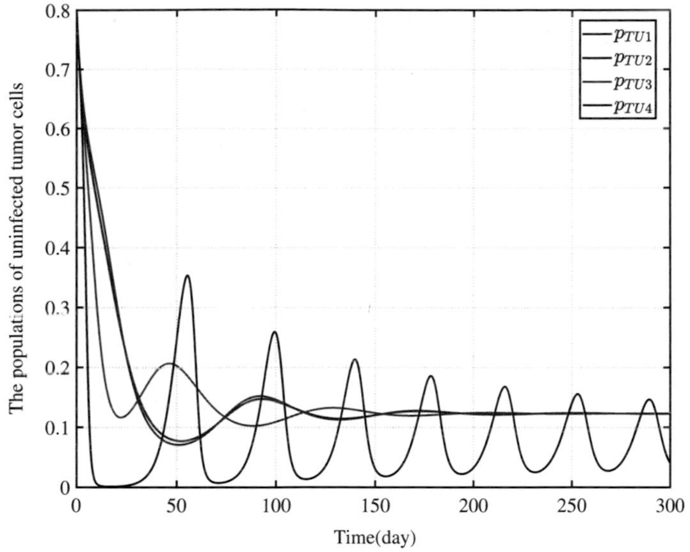

Fig. 7.6 The population evolutions of uninfected tumor cells

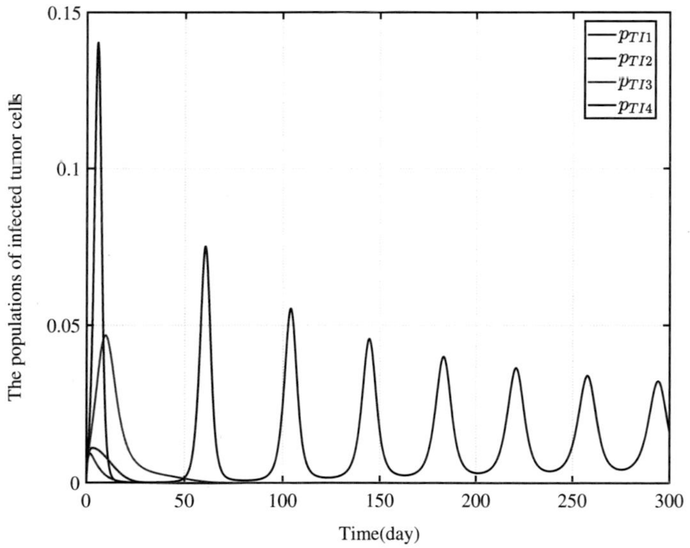

Fig. 7.7 The population evolutions of infected tumor cells

and that of virus particles which is large proportional. The immune cells are activated by the uninfected and infected tumor cells to kill tumor cells, which can be observed from Fig. 7.4. The medicine dosage of the derived approximate optimal therapeutic strategy and that of initial strategy are compared in Fig. 7.5. From Fig. 7.5, one can derive that the dosage of the obtained strategy is obviously less than that of initial strategy. On the other hand, the input dosages of the two strategies are both constrained by the pre-designed thresholds. This is of great practical significance since excess medicines may well threaten the health of patients and cause a huge overhead. Furthermore, it can be observed that the medicine dosage regulation frequency steps down when the clinic data becomes better, which means that with the aid of medicine regulation mechanism, the medicine dosage can be regulated timely and necessarily. Figures 7.6 and 7.7 present the population curves of the cells and viruses under the derived strategy with different burst sizes of viruses, that is, $\kappa = 2, 5$. This verified that the obtained therapeutic strategy can effectively kill tumor cells with oncolytic viruses of different burst out sizes. However, when the parameter κ is large enough, it may cause an oscillation. When the innate immune response is considered, the tumor-virus-immune system becomes very complicated. Though the viruses with large κ try their best to produce more replicas and infect more tumor cells, the reduction of tumor cells inactivate the immune response in the meanwhile. The viruses dominate the dynamics and the warfare between tumor cells and viruses can last a long time such that the oscillation occurs repeatedly. The oncolytic virus has the ability to effectively kill the tumor cells, while the immune response can reduce the killing-efficiency of the viruses and block their infections. Furthermore, the activated immune response can eliminate tumor cells as well. Thus there exists a subtle balance between the viruses and the immune cells which demands a further investigation.

7.6 Conclusion

Medicine regulation mechanism has been designed such that the constrained therapeutic strategy based on virotherapy can be obtained to eliminate tumor cells, guaranteeing that the medicine dosage can be regulated timely and necessarily. Firstly, a mathematical model is utilized to describe the relations among the uninfected tumor cells, infected tumor cells, oncolytic viruses and immune cells. Meanwhile, as the simplified version of the tumor-virus-immune model, the non-quadratic function is proposed to formulate the value function to acquire HJBE. Secondly, to address the optimal therapeutic strategy, single-critic architecture has been designed to seek the approximate solution of the HJBE through ADP. Finally, the simulation results has verified the effectiveness of the proposed method. Furthermore, nonzero-sum optimal control based on differential games will be a edge of the new frontier in therapy of tumor treatment, cardiovascular, orthodontic treatment, osteoporosis and cerebrovascular diseases.

References

1. Andtbacka RH, Kaufman HL, ..., Coffin RS (2015) Talimogene laherparepvec improves durable response rate in patients with advanced melanoma. J Clin Oncol 33(25):2780–2788
2. Wodarz D (2001) Viruses as antitumor weapons: defining conditions for tumor remission. Can Res 61(8):3501–3507
3. Wu JT, Byrne HM, Kirn DH, Wein LM (2001) Modeling and analysis of a virus that replicates selectively in tumor cells. Bull Math Biol 63(4):731–768
4. Wein LM, Wu JT, Kirn DH (2003) Validation and analysis of a mathematical model of a replication-competent oncolytic virus for cancer treatment: Implications for virus design and delivery. Can Res 63(6):1317–1324
5. Wodarz D (2003) Gene therapy for killing p53-negative cancer cells: use of replicating versus nonreplicating agents. Hum Gene Ther 14(2):153–159
6. Bajzer Z, Carr T, Josić K, Russell SJ, Dingli D (2008) Modeling of cancer virotherapy with recombinant measles viruses. J Theor Biol 252(1):109–12
7. Friedman A, Tian JP, Fulci G, Chiocca EA, Wang J (2006) Glioma virotherapy: effects of innate immune suppression and increased viral replication capacity. Can Res 66(4):2314–2319
8. Phan TA, Tian JP (2017) The role of the innate immune system in oncolytic virotherapy. Comput Math Methods Med 2017:6587258
9. Sun J, Zhang H, Yan Y, Xu S, Fan X (2023) Optimal regulation strategy for nonzero-sum games of the immune system using adaptive dynamic programming. IEEE Trans Cybernet 53(3):1475–1484
10. Zhao D, Wen G, Wu ZG, Lv Y, Zhou J (2023) Resilient consensus of multi-agent systems under collusive attacks on communication links. IEEE Trans Cybernet. https://doi.org/10.1109/TCYB.2022.3201909.
11. Zou B, Jiang H, Xu C, Xu J, You X, Tang YY (2023) Learning performance of weighted distributed learning with support vector machines. IEEE Trans Cybernet. https://doi.org/10.1109/TCYB.2021.3131424.
12. Zhang K, Jiang B, Ding SX, Zhou D (2022) Robust asymptotic fault estimation of discrete-time interconnected systems with sensor faults. IEEE Trans Cybernet 52(3):1691–1700
13. Littman ML (2015) Reinforcement learning improves behaviour from evaluative feedback. Nature 521(7553):445–451
14. Al-Dabooni S, Wunsch DC (2020) An improved n-step value gradient learning adaptive dynamic programming algorithm for online learning. IEEE Trans Neural Netw Learn Syst 31(4):1155–1169
15. Kamalapurkar R, Dinh H, Bhasin S, Dixon WE (2015) Approximate optimal trajectory tracking for continuous-time nonlinear systems. Automatica 51:40–48
16. Gao W, Jiang Z (2018) Learning-based adaptive optimal tracking control of strict-feedback nonlinear systems. IEEE Trans Neural Netw Learn Syst 29(6):2614–2624
17. Cui L, Xie X, Wang X, Luo Y, Liu J (2019) Event-triggered single-network ADP method for constrained optimal tracking control of continuous-time non-linear systems. Appl Math Comput 352:220–234
18. Zhong X, He H (2020) GrHDP solution for optimal consensus control of multiagent discrete-time systems. IEEE Trans Syst Man Cybernet: Syst 50(7):2362–2374
19. Zhang H, Zhang J, Yang G-H, Luo Y (2015) Leader-based optimal coordination control for the consensus problem of multiagent differential games via fuzzy adaptive dynamic programming. IEEE Trans Fuzzy Syst 23(1):152–163
20. Zheng X, Li H, Ahn C, Yao D (2023) NN-based fixed-time attitude tracking control for multiple unmanned aerial vehicles with nonlinear faults. IEEE Trans Aerosp Electron Syst. https://doi.org/10.1109/TAES.2022.3205566
21. Song R, Wei Q, Zhang H, Lewis FL (2021) Discrete-time non-zero-sum games with completely unknown dynamics. IEEE Trans Cybernet 51(6):2929–2943

22. Liu P, Sun J, Zhang H, Xu S, Liu Y (2023) Combination therapy-based adaptive control for organism using medicine dosage regulation mechanism. IEEE Trans Cybernet. https://doi.org/10.1109/TCYB.2022.3196003
23. Zhang Z, Xu J, Fu M (2022) Q-Learning for feedback nash strategy of finite-horizon nonzero-sum difference games. IEEE Trans Cybernet 52(9):9170–9178
24. Sun J, Zhang H, Wang Y, Sun S (2022) Fault-tolerant control for stochastic switched IT2 fuzzy uncertain time-delayed nonlinear systems. IEEE Trans Cybernet 52(2):1335–1346
25. Luo B, Liu D, Wu HN, Wang D, Lewis FL (2017) Policy gradient adaptive dynamic programming for data-based optimal control. IEEE Trans Cybernet 47(10):3341–3354
26. Zhang K, Jiang B, Chen M, Yan XG (2021) Distributed Fault estimation and fault-tolerant control of interconnected systems. IEEE Trans Cybernet 51(3):1230–1240
27. Zhang J, Zhang H, Feng T (2018) Distributed optimal consensus control for nonlinear multiagent system with unknown dynamic. IEEE Trans Neural Netw Learn Syst 29(8):3339–3348
28. Wang D (2020) Intelligent critic control with robustness guarantee of disturbed nonlinear plants. IEEE Trans Cybernet 50(6):2740–2748
29. Vamvoudakis KG, Miranda MF, Hespanha JP (2016) Asymptotically stable adaptive-optimal control algorithm with saturating actuators and relaxed persistence of excitation. IEEE Trans Neural Netw Learn Syst 27(11):2386–2398
30. Yang D, Li T, Xie X, Zhang H (2020) Event-triggered integral sliding-mode control for nonlinear constrained-input systems with disturbances via adaptive dynamic programming. IEEE Trans Syst Man Cybernet: Syst 50(11):4086–4096
31. Dong L, Zhong X, Sun C, He H (2017) Adaptive event-triggered control based on heuristic dynamic programming for nonlinear discrete-time systems. IEEE Trans Neural Netw Learn Syst 28(7):1594–1605
32. Ghasempour T, Nicholson GL, Kirkwood D, Fujiyama T, Heydecker B (2020) Distributed approximate dynamic control for traffic management of busy railway networks. IEEE Trans Intell Transp Syst 21(9):3788–3798
33. Chen N, Li B, Luo B, Gui W, Yang C (2023) Event-triggered optimal control for temperature field of roller kiln based on adaptive dynamic programming. IEEE Trans Cybernet 53(5):2805–2817
34. Dhebar Y, Deb K, Nageshrao S, Zhu L, Filev D (2023) Toward interpretable-ai policies using evolutionary nonlinear decision trees for discrete-action systems. IEEE Trans Cybernet. https://doi.org/10.1109/TCYB.2022.3180664.
35. Zhao J, Wang T, Pedrycz W, Wang W (2021) Granular prediction and dynamic scheduling based on adaptive dynamic programming for the blast furnace gas system. IEEE Trans Cybernet 51(4):2201–2214
36. Tian JP (2011) The replicability of oncolytic virus: defining conditions in tumor virotherapy. Math Biosci Eng 8(3):841–860
37. Al-Tuwairqi SM, Al-Johani NO, Simbawa EA (2020) Modeling dynamics of cancer virotherapy with immune response. Adv Differ Equ 2020:438
38. Yang X, He H (2020) Event-triggered robust stabilization of nonlinear input-constrained systems using single network adaptive critic designs. IEEE Trans Syst Man Cybernet: Syst 50(9):3145–3157
39. Vamvoudakis KG, Lewis FL (2010) Online actor-critic algorithm to solve the continuous-time infinite horizon optimal control problem. Automatica 46(5):878–888
40. Kuznetsov V, Makalkin I, Taylor M, Perelson A (1994) Nonlinear dynamics of immunogenic tumors: parameter estimation and global bifurcation analysis. Bull Math Biol 56(2):295–321
41. Kerbel RS, Bertolini F, Man S, Hicklin DA, Emmenegger U, Shaked Y (2006) Antiangiogenic drugs as broadly effective chemosensitizing agents. Angiogenesis, pp 195–212 (2006)

Printed in the United States
by Baker & Taylor Publisher Services